Contents

vi

Chapter 14 Modification of structure for secondary components 140

Chapter 15 Commercial and industrial service requirements 150

Chapter 16 Fire control in commercial and industrial buildings 165

Chapter 17 The control of commercial and industrial building construction 175

Part C External works

Chapter 18 Site design 184

Chapter 19   Excavations and changes of level   194

Chapter 20   Estate roads and footpaths   207

Chapter 21   Planning for vehicles and pedestrians in estates   221

Chapter 22   External services   235

viii

Chapter 28   Quality control methods   293

Chapter 29   National standards   302

Preface

This book, as indicated by its title, is primarily intended for second-year students of building technology following either the Diploma or the Certificate course in Building Studies of the Technicians Education Council. A great deal of the unit syllabus requires student research and activity: this book will supply a lot of the answers sought and, where it has not been possible to give the depth of treatment which a particular line of research may require, the references and bibliography will lead the way to further, more detailed information.

The fact that it has been written primarily for a specific student population does not preclude the usefulness of the book to students in other disciplines or, indeed, to students in other years of the same TEC discipline. Thus, I trust, it will find its way into the hands of students of architecture and building surveying who should find the review of building technologies contained within it helpful in their particular studies.

No book like this is the product of one person working in isolation and I should like to thank all those who, both consciously and unconsciously, have helped me in its preparation. To the authors of the many text books quoted in the bibliography and others not listed, which have influenced me in some way, I should like to express my appreciation of the work I know they must have put in. To my colleagues in the Building Department of the Southend College of Technology I should also like to record my thanks for assistance given, when asked, in unstinting amounts. My grateful thanks too to Gill Barritt who patiently and neatly typed my manuscript and Colin Bassett for the careful checking and direction given during the compilation of the work, to the publishers and their staff for their skilful treatment of all the manuscripts prepared in this series and finally to my wife whose constant support and encouragement have been invaluable in completing the book.

Acknowledgements

We are indebted to Schuco UK for permission to reproduce our Fig. 8.3.

Part A

Domestic construction

Chapter 1

Forms of construction

1.1 The development of constructional forms

In the past, when the term 'handling' possessed only its literal meaning of 'used by hand', all materials of which buildings were constructed had to be of sizes which a man could manage to lift, carry and fix. Thus, the traditional forms of construction are derived from the use of small building elements such as the brick. This is a perfect example of ergonomic design, in that an individual brick is the right size to fit a hand, but not too heavy to manipulate accurately and continuously, yet its size is such that the rate of construction is satisfactorily commensurate with the time taken and effort expended. Its disadvantage for modern building practice is that it is labour-intensive in an age when there is increasing movement towards the mechanical 'handling' of larger components and more off-site working.

Not all old buildings are of brick; some are built in stone and others of timber. The sizes of the pieces of stone or lengths of wood were also such as could be handled by one or two men. The choice of material was dictated by another factor: transport. When the only way to get the material to the site was to either carry it or use a horse and cart along rough tracks, the favoured material was the nearest one. To alleviate this problem with timber buildings, the posts and beams were often cut and shaped in the forest to save transporting waste timber (thirteenth-century prefabrication!). Modern vehicles and roads have changed this situation, making it feasible to use the most suitable building material, wherever it originates. It should be noted that in some parts of the country, planning authorities have imposed controls restricting the use of facing materials from other areas to preserve the local building character.

Brick, stone and timber are still used to a very large extent for domestic buildings throughout the country in the traditional manner, but only where the work is limited in extent. Modern building techniques, used in large contracts, have demanded a modification of the methods employed to utilise the raw materials and equipment, now available, to achieve economic production and building costs which are within the reach of the average house buyer.

1.2 Traditional building in brick and stone

Traditional methods, as already mentioned, employ small, easily handled building units of local material; hence, brick buildings occupy those areas of the country where clay abounds and stone in those parts where this material exists near the surface. In certain areas, specialised forms of construction developed, such as the flint walls of Norfolk and Suffolk houses.

These small units are piled one on top of another to form a wall which will both enclose the desired space and support the building loads of upper floors and roof. This form is known as a solid structure and the loads are distributed throughout the walls. It is important, therefore, that the small units of which the wall is composed act together to produce uniform strength in the structure. For this reason, bricks are laid in the familiar patterns of English, Flemish and Stretcher bond, and stones are similarly bonded but, because of the variation in size, due to the nature of the material, bonding does not always follow a regular pattern.

Originally, walls were built solid throughout their thickness but this, in many cases, did not prove satisfactory because rainwater could penetrate by capillary action. To solve this problem, walls are now built with a cavity separating an outer leaf from the inner leaf. Not only does this keep the inside of the building dry, it also makes it easier to use dissimilar materials for the two leaves. As a result, it is possible to select one material for its appearance and weathering properties for the outer leaf, and another which possesses good load-bearing characteristics for the inner leaf (see Fig. 1.1).

To complete the building, floors and a roof are required. In traditional construction these were of timber. No matter whether stone or brick clay is the predominant naturally-occurring material, trees grow everywhere, and in the past more prolifically than they do now. Consequently this aspect of traditional construction tends to be universal.

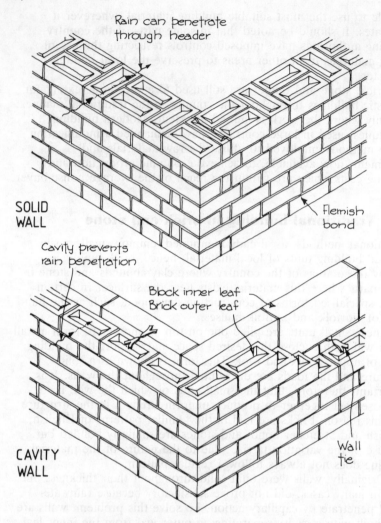

Fig. 1.1 Walls in brick

Ground floors were of wooden joists, sometimes laid directly into the soil, covered with boards butted together and nailed down. These did not last long because the timber soon decayed. Modern practice is to leave an air space below the joists which are supported on sleeper walls of brick. Alternatives to timber were beaten earth (really no floor at all), stone paving or brick paving. The first of these is a very low standard and the solid pavings laid directly on to the soil were cold, hard and frequently damp.

Upper floors in houses for single occupancy, i.e. one family per house, are still constructed of timber, in a manner very similar to long-standing tradition, with rectangular joists spanning between the walls and carrying a flooring of boards. Timber is still used for this purpose because none of the newer materials are more economical for the loads and spans encountered in normal houses.

Buildings of multiple occupancy, such as flats, must have fire-resisting upper floors and, as is explained in the next section, the most common material now used is concrete.

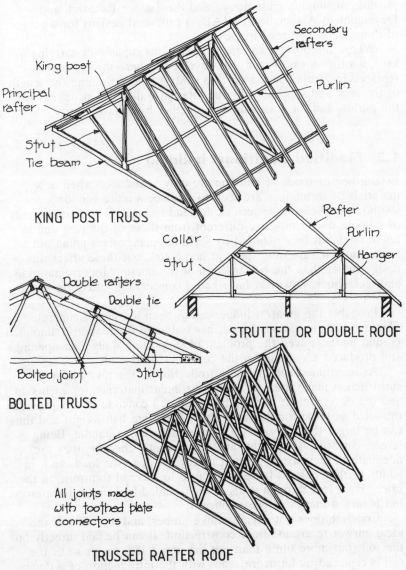

Fig. 1.2 Roof frames

The traditional form of roof construction was to frame up trusses which rested on the outer walls and carried purlins which, in turn, supported the rafters on which the roof covering was laid (see Fig. 1.2). One of the oldest forms of roof frame is the King Post Truss; it required large-section timbers and a lot of labour but produced a timber structure in which all the joints are in compression; tensile joints in timber present problems. With the advent of bolts and timber connectors (see Ch. 5), it became possible to simplify truss design and the Timber Research and Development Association (TRADA) produced designs for a range of domestic bolted roof trusses.

Where the house had internal partitions capable of carrying a load, another form of roof was developed where the truss was replaced by struts bearing on to a central binder running the length of the roof and supported by the internal walls. The struts carried the purlins and the roof was tied together by collars.

1.3 Modified traditional building

Established methods of building are only abandoned when new materials or techniques are seen to produce a more economic solution or when the pattern of demand changes. The requirements of a house owner now are different from those of the past due to few households being staffed by servants, many wives going out to work, more leisure time spent at home etc., but these affect the design rather than the construction of the house. Modern materials and techniques, however, have had a considerable effect on house construction.

Probably the greatest influence has been the use of concrete. This material, in various forms, has stabilised foundations, simplified ground floors, solved the problem of supporting walls over openings and produced economies in the construction of walls.

As mentioned earlier, the introduction of the cavity in wall construction made it easier to use dissimilar materials for each skin and the development of autoclaved aerated concrete blocks provided the ideal material for the inner skin. These are lightweight and thus can be larger than a brick yet still remain easy to handle. Being larger, construction is faster and consequently cheaper; they are adequate in their load-bearing capacity for domestic loads and, in addition, they possess better resistance to thermal transmission than brick. Heat loss through walls never troubled builders in the past, but is now a matter of great concern.

Concrete does not decay, unlike timber, and therefore is the ideal answer to ground-floor construction. It can be laid directly on the soil (but more often than not the upper 150 mm or so of the soil is replaced by hardcore) and, with the introduction of a damp-proof membrane, any ground moisture which does penetrate the

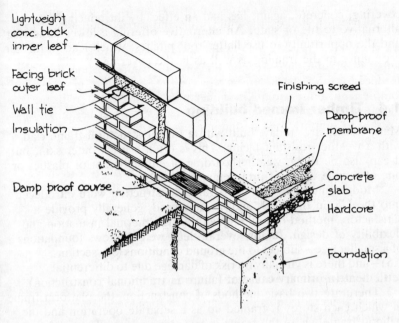

Lightweight conc. block inner leaf →

Facing brick outer leaf →

Wall tie →

Insulation →

Damp proof course →

Finishing screed

Damp-proof membrane

Concrete slab

Hardcore

Foundation

Fig. 1.3 Solid ground floor and cavity wall

concrete is prevented from reaching the finished surface (see Fig. 1.3).

Another property of concrete is its high degree of fire resistance and, for this reason, it is usually chosen for the floors of flats. A further advantage of this use is that, with the right density of concrete in the floor, a sound insulating structure is also provided. Upper floors in single-occupancy houses, where a high degree of fire resistance is not demanded, are still constructed of timber, but it is now stress graded, i.e. each piece is examined for its load-carrying quality (see Ch. 5). The assurance of stress grading means that greater reliance can be placed on the material and consequently smaller sections can be used for the joists. The flooring itself has also been changed from the traditional softwood boards around 150 mm wide to particle board panels 1200 mm square. Apart from greater economy, this change produces a more stable floor, free from the undulations which previously occurred due to the softwood boards curling, and one much more suited to receive the sheet flooring materials available at the present time.

The greatest change in roof-framing methods is the prefabricated trussed rafter (see Fig. 1.2). This has been made possible by the invention of the toothed plate connector (see Fig. 5.1), and the development of equipment by which these frames can be made, plus the greatly improved transport facilities which make it feasible to prefabricate the frames away from the site. With regard to the roof covering, concrete, again, has had an effect by introducing an alternative to tile or slate. An alternative offering a lower basic cost and the opportunity to use flatter roof pitches, which lead to further economies in the framing as well as in the covering.

1.4 Timber-framed building

Many houses are, today, built with a timber structural frame clad with a weather-resistant facing. This is often a brick outer skin, but it can also be tiling, weatherboarding of either timber or plastics or one of a wide range of proprietary materials.

Modern methods of timber-frame construction were introduced into the UK in the 1960s and, as now used, generally provide a satisfactory method of building with a high thermal insulation and flexibility of design. Its low overall dead weight allows foundations of a smaller size – subject to the ground conditions (see section 1.5) – and there is virtually no risk of damage due to differential settlement (a primary cause of failure in traditional construction).

There are two basic methods of construction–the platform frame in which each storey is framed up as a separate operation and the balloon frame where the frame is two storeys high. The platform frame is the method commonly adopted in this country.

Construction of the frame is from stress-graded timber either 100 × 50 mm or 75 × 50 mm in section, with the uprights or studs spaced at 400 or 600 mm centres (to suit a standard board size of 2400 × 1200 mm). These are fixed to head and cill members with a simple butted and nailed joint. The individual frames may be a complete wall or separate panels, depending on whether they are to be craned or manhandled into position.

The frame is lined externally with 12.5 mm sheathing plywood, nailed to the studs, which stiffens the structure against wind loads. On to this outer face, a lining of breather paper is fixed to provide a barrier against any wind-driven rain or moisture which may penetrate the exterior cladding and also to give temporary weather protection for the timber while the external cladding work is being carried out.

Thermal insulation, in the form of mineral wool or glass-fibre quilts, is secured between the studs, which are then lined with a vapour barrier, usually polythene sheet, to prevent condensation occurring within the framework. Finally, the inside finish, usually 12.7 mm plasterboard, is applied (see Fig. 1.4).

1.5 House foundations

The purpose of foundations is safely to transfer the loads of the building to the ground. This is achieved either by spreading the

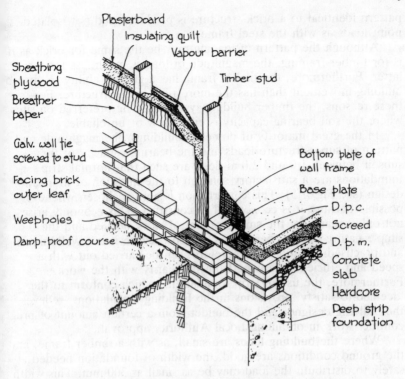

Plasterboard
Insulating quilt
Vapour barrier
Sheathing plywood
Breather paper
Timber stud
Galv. wall tie screwed to stud
Facing brick outer leaf
Weepholes
Damp-proof course
Bottom plate of wall frame
Base plate
D.p.c.
Screed
D.p.m.
Concrete slab
Hardcore
Deep strip foundation

Fig. 1.4 Timber-framed structure with brick cladding

load – at a depth clear of the effects of changes in the weather – so that the stress placed on the soil is within its bearing capacity or by carrying the loads deep into the ground to where a greater bearing capacity exists. The choice of foundation is, therefore, closely linked to the nature of the ground and the magnitude of the loads to be carried.

The nature or pattern of the building loads will also affect the foundation type. Mostly, this is a question of whether the structure imposes a loading which is more or less uniform around its perimeter, as in the case of a load-bearing brick building, or whether the loads are concentrated at specific points, such as arises with a steel-framed building. Clearly, in the first case a continuous foundation formation is needed and in the second case foundation support at the points of load concentration is all that is required.

Very few houses are built with a structural frame and, therefore, foundation structures which receive a uniformly distributed perimeter loading are those mainly required. In this connection, timber framing, as described in the last section, differs from steel framing in that the timber structure is of framed panels, which rest on a continuous cill and foundation wall. Thus, a foundation loading pattern identical to a brick structure is produced and not isolated point loads as with the steel frame.

Although the pattern of loading may be the same for brick as it is for timber framing, the magnitude of load is much less for the latter. Furthermore, the timber frame has a greater ability to absorb building movement than has the more rigid brick structure. For these reasons, the timber building system may be preferred on sites where the soil bearing capacity is either low or unreliable.

In the great majority of domestic buildings, the magnitude and pattern of superstructure loads and the bearing capacity of the subsoil between 1.0 and 2.0 m deep are such that normal strip foundations are a satisfactory solution to the question of foundation design (see Fig. 1.5). This construction is always chosen whenever possible because it is economical to construct. This economy arises not only because of the small amount of work needed and the simple nature of the operation, but also because it is a well-known and practised method and, therefore, can be carried out with a speed and efficiency deriving from familiarity with the work. Furthermore, the use of strip foundations, which conform to the 'deemed-to-satisfy' provisions in the Building Regulations, relieves the building designer and the builder from a certain amount of pre-contract work in obtaining Local Authority approval.

Where the building loads are small, as with a timber frame, and the ground conditions are good, the width of foundation needed safely to distribute the load may be as small as 300 mm. This width is very little more than the thickness of the foundation wall and allows no working space for the bricklayer. In these circumstances the trench fill or deep strip form of foundation offers a cheaper alternative although it uses more concrete to fill the trench than to lay a strip in the bottom. Trench fill can also offer a more economic construction on sites where the ground above foundation level is loose. In this situation, the cost of the extra concrete is offset by the saving in trench timbering which would otherwise be required if men were to work in the trench.

Wherever possible, a developer will choose a site for his house or his estate which does not present problems, but planning restrictions imposed to protect our countryside from a sprawl of buildings have pressed builders into tackling the more difficult sites.

The difficulties may arise through a poor subsoil, a steeply sloping site, features such as a high water-level, trees, or the close proximity of other buildings, or a combination of any of these factors.

1.6 Foundations in poor subsoil

As the bearing capacity of the soil goes down, so the width of a

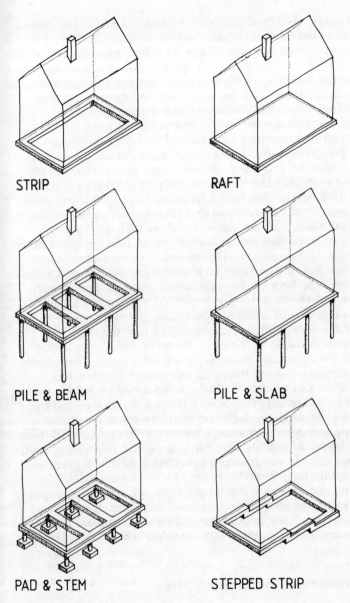

STRIP

RAFT

PILE & BEAM

PILE & SLAB

PAD & STEM

STEPPED STRIP

Fig. 1.5 Foundation types

strip foundation goes up, and consequently the depth of concrete increases to maintain the safe dispersal of stress between the top and bottom surfaces. There is quite a tight economic and practical limit to this process at which point alternative forms of foundation must be considered. The alternatives are either a raft foundation, a pad and stem foundation or a pile foundation (see Fig. 1.5).

A raft foundation achieves its aim by spreading the load, as does a strip foundation, but the raft spreads it over the whole area covered by the building, by means of a reinforced concrete slab. When conditions are so bad that a raft has to be used, a certain amount of foundation movement and slab flexure must be anticipated. For this reason, and to reduce the imposed load, a timber-framed structure may be preferred to a brick building.

If the poor subsoil overlays a firmer strata within, say, 3 m of the surface, and the contract is for one house or a very small estate, the pad and stem foundation may offer the most economic method. In this method, square holes are excavated down to the firm strata, and an unreinforced concrete pad laid (the size depending on the loads and the spacing of the pads). On these pads are constructed piers, often of large (600 mm + diam.) precast concrete pipes filled with concrete, finishing about 600 mm below ground-level. These piers are then linked by reinforced concrete ground beams, cast in trenches along the lines of the walls. The space between the piers and the sides of the excavation is backfilled with graded granular material, compacted in layers not exceeding 300 mm thick.

Pile foundations, although expensive, can, in the right circumstances, prove to be the most economic foundation system. Usually the circumstances are a very poor subsoil and a sufficiently large development to keep the piling equipment in continuous operation. Piles can be bored or driven (described by foundation engineers as replacement piles – where the soil is removed and the concrete takes its place – or displacement piles – where the precast concrete pile forces the soil aside). Bored pile foundations are most commonly found on housing sites; they are easier to construct and, although not capable of such deep penetration as driven piles, are usually adequate for the relatively light loads encountered. Furthermore, the process is less likely to disturb adjoining property than are the ground shock waves and soil displacements of driven piles. Since a lot of new housing development takes place near to existing houses, this is an important consideration.

1.7 Foundations on sloping sites

Whatever the profile of the surface of the site, the building must be raised off a level foundation, and the traditional answer to this problem on sloping sites is a stepped strip foundation (see Fig. 1.5). For most subsoils this is a satisfactory method, but there is an inherent risk which should be borne in mind, and that is the possibility of a geological fault developing. If the surface of the site slopes, there is a possibility that the planes of the sub-strata follow similar lines; if the vertical load imposed by the building on these sloping planes is too great, one layer of subsoil could slide over the one below and slip failure of the foundation can result. Where the site investigations indicate this possibility, a piled system of foundations should be adopted.

1.8 Foundations and trees

Not only must the soil below the foundations of a building be of adequate bearing capacity, it must also remain stable. In clay soils this means that the moisture content must not change. Trees draw moisture from the soil through their roots, creating a zone in which the moisture content is generally lower than normal and varying from its greatest reduction in spring and early summer, when the maximum growth occurs, to the least difference when the tree is dormant in winter. This variation can lead to the ground surface changing its level by as much as 40 mm between winter and summer. Naturally, this fluctuation, if allowed to occur beneath a building's foundations, causes serious damage.

Even if the tree is felled, the problem is not solved (unless it is felled several years before), because the cessation of the extraction of water by the roots leads to a gradual swelling of the clay as it slowly resumes its normal moisture content. A rise of as much as 150 mm can occur at the surface in the area of a felled mature tree.

Species of tree	Depth of foundation (m) for ratios of D:H of:						
	1:10	1:4	1:3	1:2	2:3	3:4	1:1
Poplar Elm Willow	Strip foundations not acceptable	2·8	2·6	2·3	2·1	1·9	1·5
All other species of tree		2·4	2·1	1·5	1·5	1·2	1·0

NOTES

D = distance from house to centre of tree
H = mature height of tree
This table applies only to isolated trees
The risk area for trees in rows is 1½ H
Table based on N H B C recommendations

Fig. 1.6 Depths of strip foundations near trees

In either situation–the presence of a tree or the recent removal of a tree–the foundations must gain a bearing at levels which will be below the root zone. For strip foundations this means taking the excavations deeper than normal, as shown in the table in Fig. 1.6. As can be seen, these depths can increase to as much as 2.8 m which can be impracticable because of ground water problems, prohibitive expense or disturbance of the natural flow of ground water, with unpredictable results to other buildings quite some distance away.

If certain problems with tree roots are apparent, a piled foundation system is much to be preferred, and if the problem is due to felled trees, the piling must be designed to resist inevitable ground heave forces which could, otherwise, pull the pile up causing damage to the building.

1.9 Foundations near other buildings

Where new foundations are to be installed near to an existing building, two dangers must be guarded against. The first is that the loads of the new foundations, when distributed in the ground, neither overload the subsoil already stressed by the existing building nor impose stress on the existing building, if it is at a lower level. The second danger is that the new foundations will remove or disturb some of the subsoil carrying the distributed load of the existing building. In either case, special foundations and construction methods must be worked out by a specialised engineer.

Chapter 2

Concepts of construction

2.1 The function of house construction

Ever since man erected his first shelter of sticks and leaves, the primary function of a building has been the exclusion of the unacceptable aspects of the climate. In the temperate areas, early man's sole concern was a shelter from the rain–whenever the weather was fine he spent his time outside (not unlike today's holiday-making tent dwellers!) In other parts of the world, protection is needed from the excessive heat of the sun and, as a result, building design and construction has always differed between countries.

By this single act of erecting a shelter within which conditions are more comfortable than outside, man modified his immediate environment. Thereby he demonstrated his superior intellect. Without exception, this has been the fundamental purpose of houses in all ages. No matter how large or small, how grand or humble, or how old or new, the basic reason why a house was build was because somebody wanted somewhere to live and, by this, he meant a location in space where the conditions were controlled to produce the very narrowly defined environment in which he could conduct his life according to his expectations. Exactly what level of expectation and how he achieved the control of the environment are matters dictated by accidents of history and do not affect the fundamental urge to erect an enclosure.

Other buildings are erected for more noble purposes of worship or commerce, but houses are built for comfort and, consequently, the whole of the shell of the structure which separates the desired internal conditions from the uncomfortable external conditions is known as the environmental envelope.

Today, we demand much more from our domestic environmental envelopes than early man expected from his covering of leaves. Not only must our houses keep out the rain, they must also exclude the wind as well as keeping in the heat we generate and keeping out the noise we also generate. In addition, they must afford protection when we lose control of our energy systems and a fire develops. While doing all this, we require the envelope to possess a resistance to the adverse effects of the weather and to exhibit an appearance which is pleasing to the eye and conforms to the standards of taste at the time the house is built.

A further complication is that the pattern of occupation of a house is multi-purpose. Unlike, say, a bank, which from when it opens in the morning to when it closes in the afternoon is solely concerned with the movement of money and has no function at other times, a house must fulfil many functions at different times. It must be a place where food can be cooked and eaten, it must also be suitable for sleeping, and most houses have to accommodate a range of activities from quiet study to noisy parties.

2.2 The elements of a house structure

Primitive man's house of sticks and leaves comprised two elements: the leaves to perform the job of deflecting the rain and the sticks to support the leaves. The leaves would not have stayed up without the sticks and the sticks on their own were useless for keeping the enclosed space dry, thus each required the other to achieve the designated building function.

This principle still applies today; every part of a building needs other parts to help it to perform its function. Take, for example, the external wall shown in Fig. 1.4. Here, the primitive man's sticks have been replaced by squared lengths of timber, but these studs still provide the support function and to assist them in this we now add sheets of plywood which stop the studs falling over. The layer of leaves is now replaced with bricks (note this is a 'leaf' of brickwork) but these are not immune from the adverse effects of ground water and so a damp-proof course is included to keep the brickwork dry. A similar damp-proof course is provided beneath the timberwork because without it, the structure would not enjoy the degree of permanence we now demand. Further refinements are selection or treatment of the brick to produce a pleasant finish, the breather paper to seal any draughts, an insulating quilt to reduce heat loss, a vapour barrier to prevent condensation forming within the structure as a result of the reduction of heat loss, and a plasterboard internal lining to give a neat appearance towards the inside of the house.

Some of the elements are indispensable, others simply add improvements, but all are present to answer the requirements of the building occupier.

Generally, we can divide the components from which a building is constructed into structural and non-structural elements. Structural elements are those which are indispensable to the primary functions and physical support of the building, and non-structural elements are those which improve upon the basic provisions. Within these main divisions can be identified elements which have a space-enclosing function, those with a support function, those with an internal division function, those with a servicing function and those with a finishing function. In many cases, functions are combined, for instance, a brick external wall both encloses space and supports the roof and there are internal walls which separate rooms and also hold up the first floor.

2.3 Structural elements

One component in a house which is never anything other than structural is the foundations. This, as already explained, has the simple but essential task of safely transferring the building loads to the ground.

Standing on the foundations is some form of construction which receives the loads of the floors and roof and transfers them to the foundations. That part of this construction which coincides with the perimeter of the building may also support the enclosing envelope and hence must also support not only the mass of that enclosing fabric, but also be capable of withstanding the wind loads which will be transferred to it. For this purpose the support construction must possess strength to carry the vertical loads and stiffness of resist the horizontal wind forces.

Strength and stiffness are also qualities required in the structural elements of the floors. Strength to carry the mass of the floor and the objects placed upon it, and stiffness to restrict the amount of bending, which always takes place, to an acceptable minimum.

In practically every house, the structural element of the roof is of timber. If it is a pitched roof, one of the structural frames illustrated in Chapter 1 and Chapter 5 would be used. If it is a flat roof, a structure of timber joists and deck, similar to a floor, is generally employed, except in a few buildings where a concrete or steel structure is used.

The only other element within a house which has to carry any significant load is the staircase, the flights of which must possess the ability to support both the people using it and anything they may be carrying, which can represent quite a high load, especially at the time of moving the house contents in or out. It is also desirable that the balustrading has sufficient strength to prevent it collapsing if someone stumbles against it.

2.4 Non-structural elements

The largest non-structural element in a house is the enclosing envelope – the outside walls and the roof covering. As already mentioned, the structural perimeter walls may also perform the non-structural enclosing function but, in many cases, these functions are separated. Even in the cavity wall, which may appear as a structural element enclosing the building, it is only the inside leaf which performs the support function; the outer leaf carries only its self-weight and can be considered as the enclosing element.

If the environmental envelope consisted solely of walling and roofing, the controlled interior of the building would be both pitch black and inaccessible. It is, therefore, necessary to perforate the walling and, occasionally, the roofing to allow the occupants to get in and out and to permit natural light to enter (we also must perforate present day envelopes for purposes of ventilation of the enclosed space, and to admit services).

Originally these perforations were left as holes which let in the light and a generous amount of ventilation, but now we fill them with doors and windows. The word 'window' is derived from the Old Norse 'Vindr' and 'Auga', which is literally 'wind-eye', indicating the draughty nature of the openings.

The next largest group of non-structural elements are the internal space dividers or partition walls. All internal walls partition or divide the building volume into smaller rooms, but not all are non-structural; some support a floor. To identify which are which in an existing building requires a careful examination and an extensive knowledge of building construction, but there are a number of clear indicators if the floor is of timber. Firstly, any load-bearing partition must be substantial, i.e. half-brick thick or 100 mm concrete blocks; secondly, it must run at right angles to the direction of the floor joists; and thirdly, it will be located at an economic spanning distance from the bearing at the other end of the joists – generally, in a house, about equidistantly from each external wall. Any partitions running parallel to the joists and any short partitions enclosing, say, a cupboard or a WC are not likely to be load-bearing even if they are thick enough (see Fig. 2.1).

The rest of the non-structural elements comprise the finishings applied to walls, floors and ceilings, and the components installed in connection with the provision of services and equipment.

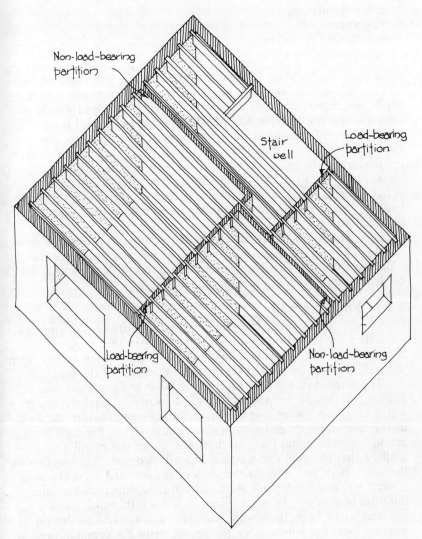

Fig. 2.1 Identification of load-bearing partitions

2.5 The enclosing envelope

As mentioned at the beginning of this chapter, the functional performance we now require of the environmental envelope are:

1. Weather exclusion
2. Durability
3. Strength and stability
4. Thermal insulation

5. Sound insulation
6. Fire resistance
7. Satisfactory appearance

In this country the attacks by the weather which the external surfaces of the building fabric must withstand are: alternate wetting and drying, wetting and freezing, penetration by rain driven by the wind, penetration by snow driven by the wind, the dead weight of settled snow, the positive and negative pressures of wind forces, temperature differences at different times of the year, and also short-term variations related to whether the sun is shining, sub-zero temperatures, the effects of ultra-violet light and the general bleaching by sunlight. All or any of these conditions can cause either a direct penetration of the envelope or a deterioration of the building fabric which, if unchecked, will eventually fail to perform its function.

The intensity of these attacks by the weather depends on whether the surface is vertical, inclined or flat and, therefore, the walls must be considered separately from the roof which must, itself, be divided into pitched roofs and flat roofs.

2.5.1 Enclosing walls

Direct rainfall has very little effect on the majority of wall surfaces since they are protected by the protruding eaves of a roof or by a coping, but if the rain is accompanied by wind, as often occurs, then the wall face will be wetted on the windward side of the building.

The traditional walling materials of brick and stone are generally porous and permit the penetration of rainwater by capillary action which continues for as long as the surface is being soaked. As soon as the rain stops, evaporation at the surface begins, the capillary action is reversed and the wall dries out. From this it will be realised that the depth to which damp will penetrate is a function of the quantity of rain and the duration of the storm. Unless the area is very dry, one brick thick solid walls are inadequate; where the duration of rain storms is measured in hours rather than days, a one-and-a-half brick thick solid wall will probably remain dry on its internal face; but where more prolonged wetting is common, an even thicker solid wall may be needed. The alternative construction of a cavity wall is now almost always adopted because the lack of continuity between the outer and inner leaf, created by the cavity, prevents the former, which may be completely saturated, from passing any of its moisture to the latter. For this reason it is important that the cavity is not bridged by anything other than a properly made wall tie (all types of which incorporate some device to prevent water running along them).

In many houses, materials other than brick or stone are used for the external face, but in each case the possibility of penetration by

driving rain must be carefully examined, particularly where any
joints occur. Joints can either be 'closed' or 'drained'. Closed joints
are those in which the gap between two adjacent components is
sealed at the surface to prevent any rain penetration at all. Drained
joints allow the rain to be driven into the joint to a pre-determined
distance, at which point an enlargement of the joint gap reduces the
pressure of the wind and further penetration ceases. The rainwater
is then caught by a baffle and conducted harmlessly away (see
Fig. 2.2).

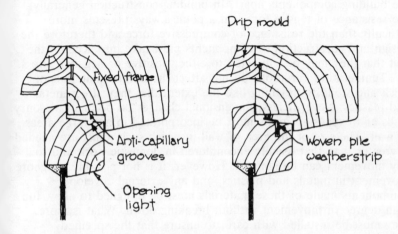

RAIN EXCLUSION IN WINDOW JOINERY

Fig. 2.2 Rain exclusion at joints

Driven rain and draughts are more likely to penetrate around
the windows and doors in the walls than the walls themselves,
largely because of the need to be able to open them. The traditional
methods for dealing with this problem are to use anti-capillary
grooves, drip moulds and throatings. Modern plastics have allowed
weather stripping to be added to the methods (see Fig. 2.2).

As well as forcing rain into every crack and crevice of the
building face, the wind can exert forces which must be allowed for
in the design and construction of the walls. The brick and stone
constructions normally encountered in domestic work do not usually
present any problems in this respect, mainly because of the inherent
strength and mass of the material, but any lightweight constructions
or components such as windows need careful consideration.

The forces to be dealt with are positive pressures on the face on
the windward side and suction (negative pressure) on the leeward
side. Positive pressures are dealt with by providing adequate

strength in the materials to ensure that they can transfer the forces to their supports and by the provision of lateral restraints in the building to accommodate these transferred forces. Given satisfactory strength and stability, the effect of the positive pressure is to press the building components more tightly together. Such is not the case with negative pressures. On the leeward side of the building, not only must the same conditions of strength and stability apply, but also strength of fixity because the suction (and internal outward pressure from leakage through the building) has a tendency to tear the building components apart. In building construction generally, the resistance of tensile forces at a point always presents more difficulty than the resistance of compressive force and therefore the design and construction of components must take into account the fact that they may be subjected to either positive or negative stress.

Temperature variations do not affect the natural materials of brick and stone to anything like the extent that they affect metals and plastics. Nevertheless, dimensional changes do occur in masonry walls and expansion joints must be incorporated in any large areas of wall. The size of the area of wall in which such joints are needed is greater than that usually encountered in domestic work and so any movement can be ignored. However, it is not possible to ignore movement in metals and plastics, and any external envelope components made of these materials must be designed to allow for a high degree of movement without breaking down. What is more, they must be installed with care, to ensure that the specified clearances are maintained.

Changes due to ultra-violet light and bleaching effects of the sun do not affect brick or stone to any appreciable extent either, but do manifest themselves with other materials found in the enclosing walls and windows. Ultra-violet light breaks down natural oils and consequently any materials used externally containing such oils will, in due time, become brittle and disintegrate. The materials commonly used which fall in this category are bitumen-based products and paints. The problem with bitumen products is more acute on roofs where they are more extensively used (see the next section) and the only satisfactory answer is to provide a cover or shield to prevent the ultra-violent light from reaching the material. In the case of paintwork, it cannot be protected since it is itself intended to be a protective coating and consequently, the practice is to accept the fact that a breakdown of the paint film will occur, and repaint every three years.

Bleaching by the sun is most noticeable in exposed timber. Many houses are finished, for example, with western red cedar boarding which, when the house is new, exhibits a richness of colour but which rapidly reverts to a dull grey, even within one year. The only answer is, as for paint, to restore the original quality by regular maintenance.

2.5.2 Pitched roofs

Roofs are pitched because the covering materials applied would let in the rain if it was flat, not because the designer considered a sloping roof looked better than a flat one. Familiarity with this building shape has persuaded us that this traditional design is to be preferred. In the drier areas of the world, flat roofs predominate, and a pitched roof would be out of place.

The traditional materials which have dictated the pitched roof are, of course, plain tiles and slates. These small roofing units must be inclined at an angle which will ensure that rain runs off the bottom edge and does not spread out to dribble over the sides, nor must it be possible for the wind to blow the rain back up under the bottom edge. The ability to keep the rain out is improved by increasing the size, thereby reducing the number of joints through which the water can flow. There are two reasons why this is not done: firstly, there is the question of handling, discussed in Chapter 1 – the size of the tile or slate must be small enough for a tiler or slater to manipulate; secondly, the larger the unit the greater the strain on the fixings when the wind blows. On the windward side the effect of wind pressure is to lift the bottom edge of the tile or slate and to lever the fixing nails out at the top. On the leeward side the negative pressure simply sucks the covering off the roof.

Failure to provide an adequate pitch angle can also result in the rain soaking into the tile rather than running off, and if a frost occurs before the tile has had time to dry out, there is a high risk of the tile flaking or delaminating. Modern concrete tiles, being more homogeneous than clay tiles, are less prone to this attack and can, therefore, be laid at lower pitches.

To allow advantage to be taken of the ability to lay concrete tiles at a lower pitch, the design of the tile has been modified to direct the rainwater more positively towards the bottom edge, by means of ridges formed down the length of the tile. In some designs, where the tile is intended for very low pitches, there are also inter-locking grooves and ridges in the overlap between the top and bottom edges of successive tiles to stop driven rain from penetrating.

Some manufacturers provide special stainless steel clips to hold the bottom edge of the tile firmly against lifting by the forces of the wind. These are more severely felt on the flatter pitches to which modern tiles can be laid.

The reason for the striving to reduce the roof pitch is cost: the flatter the roof the smaller the area of the slopes and the fewer the tiles needed to cover it. In addition, a low pitch also saves on the quantity of timber required in the supporting frame.

Because of the small size of the roofing units and the nature of the materials used on the majority of domestic roofs, movement due to temperature variation is not a problem, nor is the effect of ultra-violet light a serious matter with tiles or slates. The earliest of the concrete tiles proved not to be colour-fast but makers have now developed their products to the point where those now manufactured only lose a small amount of their original colour during their lifetime.

2.5.3 Flat roofs

This is a misnomer. No roofs are flat, they all slope (are laid to falls) so that control is maintained over the direction in which rainwater flows. The amount of fall needed depends on the particular covering used and the surface upon which it is laid. The angle of 10 degrees is taken as the dividing line between a flat roof with falls and a pitched roof.

Since the rainwater runs more slowly off a flat roof than off a pitched roof, the covering material either must be jointless or must be jointed by a method which prevents the ingress of water. The materials used are metals – lead, zinc, aluminium and copper – and bitumen-based products – asphalt and bituminous felt. There are also some proprietary products which make use of a number of the plastics materials now produced.

While fairly large in size, sheets or rolls of metal roofing are never big enough to cover a complete roof in one piece and, therefore, joints must be formed. These are worked in a variety of ways, but all function on the principle of turning the edges of the sheet up and out of the water flow. Thermal movement must also be accommodated in metal roofing, and careful design of joints and fixings must be observed to avoid ruptures occurring in the sheet.

Ultra-violet light has no effect on metal roofing but it is a serious problem with bituminous coverings. Natural asphalt is the most stable in this respect and will last a long time, but is expensive. The cheaper alternative, bituminous felt roofing, is severely attacked if left exposed, and must always be covered with light-coloured stone chippings to reflect the sun's rays, or with some other covering such as lightweight pavings (if the roof is accessible for pedestrians).

Being nearly flat, this form of roof suffers little from the force of the wind, provided that the bottom edges of the covering sheets are securely held so that they cannot be lifted. There is, however, an interesting case of the combined effects of weather causing a breakdown. On some felted roofs, the wind can gradually sweep the stone chippings to one side or corner, especially if there is a parapet to the roof. This causes no harm to the roof but it does expose the felt to the ultra-violet light of the sun which then breaks down the oil in the bitumen, leading to crazing of the felt. This allows seepage of rainwater to below the covering which, when the sun shines, evaporates, causing blistering of the felt which, now being brittle because of its loss of oils, cracks badly thus failing completely as a weather-excluding envelope.

Chapter 3

The process of construction

3.1 Integration of the construction processes

In the building of a house (and any other building), many different
operations must be performed, and many tradesmen employed to
contribute their skills and to carry out their work to achieve
satisfactory completion. These operations are not performed in
isolation: one tradesman's work is always related to that of several
others. To ensure that the whole course of building proceeds
smoothly and efficiently, these inter-related activities must be
suitably integrated by planning the work in advance.

This planning of site operations is one of the main jobs the site
foreman has to do and involves two main considerations. Firstly, he
must make sure that the right men and the right materials in
adequate quantities are in the right place at the right time and,
secondly, he must ensure that all requisite preceding work has been
completed, and none of the activities being conducted at the same
time will get in each other's way.

To achieve a smooth operation, it is often necessary for certain
tradesmen to carry out their work in separate stages. For instance,
the electrician will be brought on to the site to fix his conduits and
cables, plus the boxes into which the switches and sockets are fitted,
as soon as the structure is weathertight, but he cannot complete the
installation until after the plasterer has carried out his work. Once
the walls are finished, the electrician can return to fit the switches
and sockets and to complete the rest of the electrical system. The
weathertight stage is usually marked by the tiling of the roof and

the glazing of the windows thus, surprisingly, the electrician's work must be timed to fit in with that of the roofer, the glazier and the plasterer, and these trades must be programmed with the preceding activities of the bricklayer and the carpenter as well as the succeeding trades of the joiner and the painter.

The organisation of the site, the accommodation to be provided and the contractor's responsibilities are dealt with in Part D of this book; the following sections of this chapter deal with the integration of the various processes of construction.

3.2 Groundworks

Usually the most difficult and awkward part of a building project is the period when the groundworks are in progress. This term embraces all the site operations which take place at or below ground-level, and which comprise the building substructure of foundations and oversite concrete, the drainage system and the public utilities of water, gas and electrical services.

The difficulties arise because three other factors are added to the ever-present factor of human error. These other factors are the weather, variations in the nature of the ground, and the location of the work. Any work carried out in the open is subject to the vagaries of the weather. When digging holes and working within them, rain can seriously hamper operations, and when using wet materials such as concrete; frost, rain and strong sunlight can affect, or even stop, the progress of construction. Many surprises lie in wait for the contractor when he starts excavating a site: some are natural such as localised changes in the subsoil or an unknown spring, and some are man-made such as a forgotten well, an unrecorded mains electrical cable or archaeological remains. All can delay progress and some may change the intended method of work or design of the structure.

3.2.1 The substructure

The substructure of a house consists of the foundations, the foundation walls and the oversite concrete.

Once the contractor has established himself on the site, arranged his access point, storage area and site accommodation, the first operation is to strip oversite. By this is meant the removal of all the vegetable topsoil over the area to be occupied by the building. Topsoil is compressible and so cannot withstand building loads, but it could continue to support plant growth below the building if left, with harmful results. Disposal of the stripped topsoil is either by spreading it on the site or by carting it away, whichever the building owner directs.

Having exposed the subsoil, the building team then takes out the foundation trenches. If these are deep, or if the subsoil is liable to collapse into the trench, a support system of polling boards and struts must be installed as quickly as possible to prevent any accidents. Before placing the concrete in the trench, the bottom must be inspected by the architect and by the local council's building control officer to check that it possesses the requisite bearing capacity. If deep strip foundations are specified, the foreman must also make sure that any ducts required for drain pipes or service runs to enter the building through the concreted foundation are correctly positioned.

Placing the concrete in the trench should be done as soon as possible (usually within two or three days) to minimise any change in the subsoil condition due to exposure to the weather, and with care to maintain the quality of the concrete. If a wet mix of concrete is dropped from even a relatively low height of a metre or so, the heavier constituents in the upper part of the mix continue to descend after the lighter constituents have stopped moving. As a result, concrete thrown or tipped into a trench tends to have all the stones at the bottom, the sand in the middle and the cement at the top, and is weakened thereby. It must be placed by the use of a chute or pipe.

The level of the top of the concrete will determine the position of the house in relation to ground-level and, therefore, must be correctly set. This is achieved by driving wooden pegs into the trench bottom, prior to concreting, so that their tops are at the desired concrete level. It is then simply a matter of placing concrete until the pegs are just covered.

When the concrete has set hard, the foundation walls are built up to the level of the oversite concrete bed, and the foundation trenches are filled up. Inside the walls, the trench back-filling is with hardcore; outside the walls, selected excavated material is returned, finishing with a layer of topsoil.

Where the strip foundations are used, the allowance for the entry of drain pipes and service runs, mentioned above, usually must be made in the foundation walls. This can be achieved either by building sleeve pipes into the wall through which the service pipe can be fed at a later date, or by building 'sand courses' in the wall, i.e. laying the bricks in sand, rather than mortar, over a small area of the wall. When entry is required, the sand courses can easily be removed, the pipe positioned and the space around the pipe solidly re-built with bricks – using mortar this time. The sand course method allows greater positional tolerance for the pipe and, therefore, is the one generally favoured.

Following completion of the foundation walls and back-filling the trenches, preparations are made for the oversite concrete by spreading and compacting a layer of hardcore. This must finish 50 mm below the underside of the concrete bed and be not less than

20·08·64

100 mm thick. On this hardcore is spread a 50 mm thick bed of sand or similar fine material, referred to as 'blinding', to provide a smooth level surface.

Before placing the concrete, the foreman must check where the damp-proof membrane is to be placed in the floor. There are three positions it can occupy: under the slab, on top of the slab and under the screed, or on top of the screed. If the damp-proof membrane is to go below the concrete slab, it will be of heavy-gauge polythene sheet and laid over the blinding and up the foundation walls, just prior to concreting. Barrow runs of planks must be provided and great care taken to ensure that the polythene is not punctured while the concrete is being placed.

Finally, the concrete is placed, spread and levelled to the top of the foundation walls. The minimum thickness stated in the Building Regulations for this oversite concrete, is that it should not be less than 100 mm and some specifications require 150 mm. If the damp-proof membrane is to be located between the slab and the screed, the concrete will be finished with a steel float to give a smooth surface suitable for painting with the damp-proofing liquid used for this type of damp-proof membrane.

3.2.2 The drainage system

The problem with the integration of the drainage work (and also the installation of the external services) is the inconvenience of the trenches which interlace the site and the banks of excavated material. These can seriously hamper the movement of materials and men about the site and make the erection of scaffolding impossible. Furthermore, it is unwise to drive heavily-laden vehicles over or along the line of freshly-laid drains as damage can result.

The timing of the drainage work is frequently arranged for the latter part of the building programme; after the brickwork, concreting and roofing have been completed, all the heavy materials have been used and the scaffolding cleared away. An exception to this timing is the installation of drain runs below the building. These are always required to provide a connection to the soil and vent pipe and also for ground-floor WCs or any internally-positioned sealed gullies. Such drainage branches must be put in before the hardcore and oversite concrete is placed.

While it is helpful to delay the bulk of the drainage work until towards the end of the contract period, it must be integrated with certain following trades. There are two main areas of work which need the drains completed, the installation and testing of the sanitary fittings and the laying of external pavings, drives, turf etc. to complete the external works. If the line of the main drain runs below the footpath crossover, on its way to its connection with the sewer below the road, the timing of the drainage work must also be set to coincide with the digging up and reinstatement of the road and crossover, and the making of the sewer connection. This may have to be carried out by a contractor and not the firm building the house. The situation is different on new estates where the sewers are laid with branches to each site before the work on the roads and footpaths is started.

3.2.3 The public utilities

Water and gas services are always laid underground, main electricity cables usually are buried and telephone cables are buried wherever possible. Thus the installation of these public utilities creates the same problem as the drainage system and, for that reason, may be installed at the same time as the drains.

To avoid freezing, the underground water main must terminate not less than 600 mm inside any external wall and must be laid not less than 760 mm below ground-level (see Fig. 3.1) If flexible tube, such as soft copper or polythene, is used for the water main, it can be threaded through a sleeve pipe to the internal stopcock position, but if the main is to be in rigid pipe such as p.v.c. or galvanised steel, this must be installed at the same time as the internal drains, i.e. before the hardcore and oversite concrete is laid.

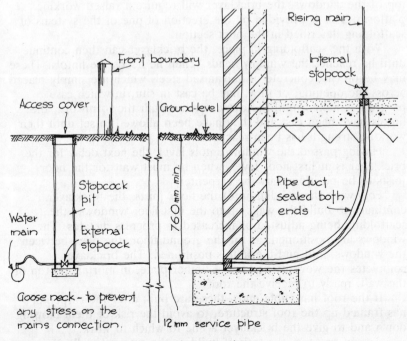

Fig. 3.1 Underground water main connection

3.3 Superstructure

In Chapter 1, two forms of construction for houses were described, solid structures of brick and block, and framed structures of timber. Each of the superstructures requires different control and integration within the building programme.

It is becoming common practice to place 465 × 900 mm insulating slabs or batts in a cavity wall as it is built, and hence the usual procedure for building the external walls of a house now is to raise the inner leaf in lightweight concrete blocks to two courses above the outer leaf, place the insulation in position, raise the outer leaf in brickwork, place the wall ties to bridge the two leaves at 900 mm centres and continue with the inner leaf. The procedure is maintained until the ground-floor window cill level is reached.

Windows are usually built in as the wall continues up and, therefore, the bricklayer needs the window frames' positioning on the walls before he can carry on. These frames are accurately located both in their relationship to the rooms they are to light and in their position in the wall thickness. Once correctly placed, they are checked for uprightness with a spirit level or plumb line and supported by a stay taken down to the oversite concrete. Bricklaying then continues up past the windows with the cavities being closed and a vertical damp-proof course built in. Before he can reach the top of the windows, the bricklayer will require a raised working platform. This is provided by the erection of one of the systems of scaffolding described in the next section.

With the scaffolding in place, the bricklayer can then continue until he reaches the window heads. Then he needs the lintols. These may be precast concrete or galvanised steel, which are simply placed across the opening, or they may be cast in situ, in which case progress is halted until the carpenter has built the formwork, the lintols have been cast and they have been allowed to set until their working strength has been reached.

Having passed the window lintols level, the next delay for the bricklayer is at first-floor level, when he must wait for the floor joists to be positioned by the carpenter.

Following the building-in of the floor joists, the bricklayer continues to build the walls up to the first-floor windows – the scaffolding being adjusted or increased as progress dictates. The windows are positioned, as for the ground floor, the piers between the windows built and the lintols positioned. The bricklayer completes the work by bedding a timber plate, in mortar, on top of the wall, ready to receive the roof.

If the roof has gable ends, these are built after the carpenter has framed up the roof structure, to avoid the risk of them blowing down and to give the bricklayer a line to which to work.

As can be seen, the task of building the external walls of a house proceeds in stages and must be integrated with the work of the carpenter and, possibly, the concretor, and requires the attendance of the scaffolder.

Once the walls are up, the carpenter then moves in to construct the roof frame or to fix the trussed rafters – if this is the system specified – followed by the roofer who completes the enclosure of the building.

One of the advantages claimed for timber-frame construction is that the roof is finished earlier in the programme, thus providing shelter for the workmen and the building interior sooner than the cavity wall method.

The wall frames may be made up on-site, if they comprise the whole side of one storey, or they may be prefabricated off-site in smaller panels. On-site fabrication often makes use of the oversite concrete for a working surface and a template of the building sizes. Each frame consists of top and bottom horizontal members, called a top plate and bottom plate or head and cill plates, linked by vertical members called studs. The frames are assembled in a horizontal position and the joints are made by butting the timbers together and nailing. As it is completed, each frame is raised and stacked out of the way.

In many building programmes, the next step is to assemble the first-floor frame on the oversite concrete 'template'. This frame consists of the floor joists, spaced at 400 mm centres, held across their ends by a header of the same size as the joists and rows of strutting between the joists to prevent them from twisting. The assembly of the floor frame is then followed by the assembly of the roof frame, on top of it. Trussed rafters are generally used and these are spaced out at 400 mm centres and nailed to timber plates laid along the edges of the floor frame. One truss is accurately plumbed and held with a temporary stay, then the tops of the rest of the trusses are set out from it and held with a runner fixed to the underside of the rafters. Wind-bracing timbers are fixed, also to the underside of the rafters, running diagonally from eaves to ridge to hold the whole structure square and true.

With all the frames made up, the process of assembling the structure begins. For this purpose, a mobile crane is generally used. Firstly, the roof frame is lifted out of the way and then the floor frame is raised. With the oversite clear, a base plate is accurately bedded and levelled on top of a damp-proof course on the foundation walls and the ground-storey frames are placed in position, plumbed, braced and fixed at the corner junction. Then the first-floor frame is lowered into place and secured. Next the crane lifts the upper-storey frames on to the edge of the floor frame – these are also plumbed, braced and joined at the angles. Finally, the roof frame is picked up in one piece, placed on top of the upper-storey frames and the roof plate nailed to the top plate of the storey frames. Assembly of the structural frame is usually completed in a few hours and, with the roof frame in position, the crane can be taken away.

As soon as the whole framework is assembled, it is clad in sheathing plywood, nailed to the studs and plates, to provide lateral stability to the structure in readiness for the loads of the roof and internal constructions.

The envelope is completed by felting and tiling the roof, and by cladding the walls. Brick cladding is described in Chapter 1 and illustrated in Fig. 1.4. It is frequently selected as the finish because brick is a very stable and durable material, and because it gives the traditional appearance many house owners require, with little maintenance.

There are other materials which can be employed to clad the frame which have a lower initial cost than brick, but usually are more expensive to maintain. The three principal materials used for houses are tile hanging, boarding and rendering.

Tile hanging, i.e. plain tiles hung on battens nailed to the timber structure (see Figs 3.2 and 5.4), is a traditional technique suited to a timber structure because it will accept movement of the structure without any damage. It is also as resistant to rain and snow as brick. The main disadvantage of tile hanging is that it is fairly easily damaged either by dislodgement of the tiles in a strong wind or by accidental impact.

Softwood boarding is available in two forms: fine sawn feather-edged, suitable for treating with timber preservative or bitumen; and shiplap, which is a planed section and thus can be painted (see Fig. 3.2). This is a cladding method which is quick and easy to carry out, low in its initial cost, but very high in maintenance costs. Like tile hanging, it also accommodates structural movement with ease. There are also proprietary plastic weatherboards, produced in p.v.c. and glass-reinforced plastics, which simulate timber boarding, but have practically no maintenance. Some incorporate a thermal insulating core.

Rendering is the process of covering a suitable surface with coats of a mixture of sand and cement plus lime. It is the term used for external finishes. Plaster is the term used for similar finishes internally. BS 5262 : 1976 gives a code of practice for external rendered finishes and recommends various proportions of mixes for different types of backgrounds, i.e. the surface to which the rendering is applied. In timber-framed structures the background is invariably expanded metal lathing (e.m.l.). Expanded metal lathing looks rather like heavy wire netting but is actually made from a sheet of steel which is covered in short slits and then stretched sideways so that the slits become holes. After being expanded, the sheet is galvanised to prevent it rusting.

The rendering adheres to the e.m.l by a mechanical grip

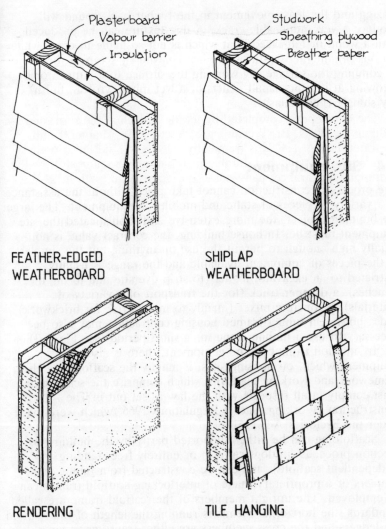

FEATHER-EDGED WEATHERBOARD

SHIPLAP WEATHERBOARD

RENDERING

TILE HANGING

Fig. 3.2 Claddings for domestic timber frames

achieved by passing through and surrounding the mesh. To allow this to happen, the e.m.l must be spaced away from the plywood and breather paper surface, on battens (see Fig. 3.2). To achieve a satisfactory surface on e.m.l, it is recommended that three coats are applied, the first to grip the lath, the second as an undercoat to give substance to the rendering, and the third as a finishing coat to which a surface treatment is applied. A smooth, close, dense surface can be produced by working the final coat with a steel float (the rectangular trowel the plasterer uses), but this is very prone to crazing and the least movement in the timber background will produce a very noticeable crack. A fine texture can be produced with a wood or felt-faced float which is not so liable to craze while various ornamented textures can be achieved by scraping, brushing or combing the surface, by working the surface to a pattern or by throwing at it either small pellets of a wet mix (Tyrolean finish) or dry shingle (pebbledash).

3.4 Site equipment

The process of construction cannot take place without the assistance of a variety of pieces of static and mobile site equipment. The larger the building project, the more extensive and sophisticated the site equipment must be. In house building, the contract value is not usually high enough to permit the use of anything but the simplest of the pieces of equipment available and the range is usually restricted to an excavator/loader (to strip oversite and to dig the trenches), a dumper truck (for the transport of a variety of materials), a concrete mixer (mainly to mix mortar for brickwork) and a hoist. On timber-framed housing contracts it may also be necessary to use a mobile crane for a short period (see Fig. 3.3).

In addition to mobile site equipment, there is also static equipment which, on housing sites, is mainly the scaffolding. The framework and working platforms which comprise the scaffolding must comply in all respects with the law as set out in The Construction (Working Places) Regulations 1966, which were made under the Factories Act 1961.

Scaffolding can be either supported partly by the building as erection proceeds (putlog scaffold) or entirely free-standing (independent scaffold), and can be constructed from tubing and couplers or a proprietary make of interlocking scaffold frames can be employed. The upright members of the scaffold frame are called standards, the horizontal members running the length of the scaffold are ledgers and the cross members are either transoms, in an independent scaffold, or putlogs, in putlog scaffold. The frame must also be provided with diagonal braces to hold it rigid and, at each working level, a guard rail. The feet of the standards must be fitted with base plates which must be placed on a timber sole plate where the ground is soft. The working platform is composed of scaffold boards placed close together and butted end to end (not overlapped) and it must be fitted with a toe or guard board which is a scaffold board clipped, on edge, to the standards (see Fig. 3.4).

A typical proprietary frame is also shown in Fig. 3.4 and is based on an 'H' frame of tubing. Each frame fits into sockets on the

35

frame below. When assembled, they form towers which can be used in that form or linked together to make a continuous or face scaffold.

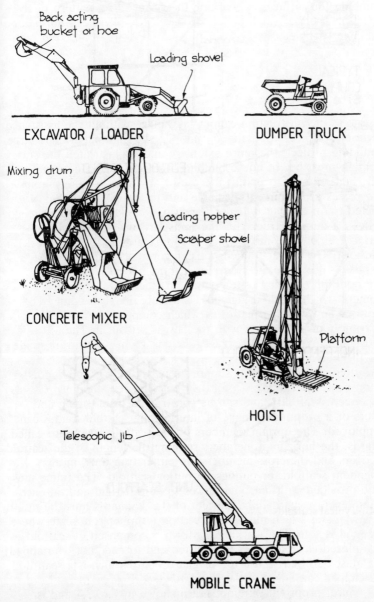

Fig. 3.3 Site plant

36

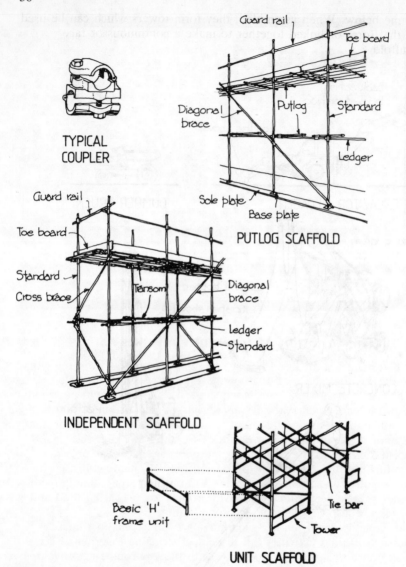

Fig. 3.4 Typical scaffolds

Chapter 4

Internal domestic construction

4.1 The function of internal walls

When man first enclosed a space to separate himself from the worst of the weather, he used the volume thus provided as a single cell, in which all his activities were conducted. This arrangement persisted until the early medieval days, as the large, single-roomed Norman manor house bears witness, but it was not long before the lord and lady of the manor found it desirable to withdraw from the boisterous life in the great hall, and another space was provided for this purpose, called a withdrawing room (a name still used by some house owners but contracted to drawing room). Similarly, other spaces were added to provide rooms for preparing food and drink and also for sleeping and for worship.

Today we still require separate spaces or rooms in which to pursue different activities but instead of stringing together a series of spaces, we enclose a total volume with the environmental envelope and then divide it by light partitions. In some cases these partitions may afford a structural support to an upper floor (see Ch. 2) but their principal function is to divide the enclosed space in such a manner that the rooms thus separated are suitable for their intended purpose.

The achievement of satisfactory partitioning poses three secondary requirements which, in a house, are readily met. Firstly, the wall must possess sufficient strength to support itself within the storey height and to resist any lateral loads from objects or occupants leaning against it. Some partitions must also be strong enough to carry the weight of shelves, cupboards or sanitary fittings fixed to them. Secondly, partitions separating spaces occupied by different families must be sound-insulating and, thirdly, those between occupied areas and escape routes must be fire resisting.

There are four types of internal partition commonly used in housing: brick walls, block walls, prefabricated slab partitions and timber stud partitions.

4.2 Brick partitions

Brick partitions are usually built half-brick thick in stretcher bond, that is, the bricks are laid lengthwise in the wall so that their long or stretcher faces are exposed and each brick is centred over the joint between two bricks below. The mortar joints are not carefully finished, as in external face work: any excess mortar protruding from the joint is struck off, but any irregularities of the joint face are left, as these assist the plaster to gain a grip.

Brick partitions have the greatest strength, fire resistance and sound insulation of any partitioning method, but they also tend to be the most expensive and, consequently, other methods are preferred unless the superior qualities of brickwork are needed. The greater strength of brickwork derives from the density and inherent strength of the individual bricks and also from the stronger mortar in which they can be laid. Its fire-resistant quality stems from the same sources. Its sound-insulating quality is due to the mass of the partition being greater than that of any other method and, therefore, more capable of absorbing sound energy.

One of the main disadvantages of brickwork partitioning is the mass mentioned in the last paragraph. Partitions, especially upper-storey partitions, must be supported by the structure and, therefore, are generally required to achieve their function with the minimum imposed load. Brick partitions are generally too heavy to be carried by a normal domestic upper floor and, therefore, a beam must be specially provided, or the partition arranged to stand on a similar wall below, which is carried down to its own foundation. Usually, where brick partitions are selected, it is for fire-resistance or sound-insulation purposes and, in these cases, the partition runs up through the full height of the building, e.g. the party wall between the two halves of a semi-detached house.

The high stability of brickwork exposed to fire makes it a fairly common choice for the structure of chimney breasts and the part of a partition which encloses a flue, even where the rest of the partition adjoining the breast or chimney may be constructed by one of the other methods.

The safe thickness of any partition depends on its size and the degree of fixity around the edges. In the process of building a house, the internal, non-load-bearing partitions are not usually built at the same time as the external walls. The traditional method of providing edge-fixing for brick and block partitions is to leave space at intervals in the inner-leaf of the cavity wall into which the bricks or blocks can be inserted when the partition is built. This is known as 'block bonding'. Given adequate restraint at the edges, it is generally taken that the thickness of a non-load-bearing partition should be not less than one-fortieth of its length between buttressing walls or its height between floors, whichever is the greater. There is no requirement laid down in the Building Regulations of a minimum thickness for a non-structural partition, but clearly a minimum exists, and the one-fortieth rule is found in the London Building Acts.

Since the thickness of a brick partition is fixed by the size of a brick, the rule given above can be inverted to calculate the maximum size of partition; in this case, it works out to be 4500 mm for a 112.5 mm partition (5500 m if the thickness of plaster on each face is included). This must refer to the length of the wall, since storey heights of this dimension are never met in houses, and it is an exceptional house plan that has an internal partition this long without any other partitions connecting to it.

4.3 Block partitions

Building blocks are walling units which are larger than bricks but in which the height does not exceed the length nor six times the thickness, to distinguish them from slabs or panels. The materials used in the manufacture of partition blocks are concrete or clay.

Concrete blocks are either dense or lightweight. The dense concrete blocks are only used for non-structural partitions where a high level of sound insulation is required. Lightweight blocks are produced by either using a lightweight aggregate bonded with cement or aerating a mix of sand, lime and cement to give a cellular block. There are a number of lightweight aggregates, but the blocks most commonly available use either expanded clay nodules, which look rather like hollow clay beads, or expanded vermiculite. The block most commonly seen on domestic sites is the aerated block, often, and quite incorrectly, referred to as a breeze block.

Autoclaved aerated concrete blocks are made from a carefully graded slurry of sand, lime and cement to which a small amount of aluminium powder is added. The powder acts with the alkaline slurry to produce hydrogen gas which causes the mix to foam. This is allowed to set in a large mass which, at a suitable stage in the setting process, is cut into block-sized pieces. These blocks are then placed in an autoclave where they are steam cured at high pressure and temperature.

Concrete blocks are made in a variety of sizes, the most readily available being 400 × 200 and 450 × 225 mm in thicknesses of 75, 100, 140 and 215 mm. The face sizes given are the format sizes and include the thickness of the mortar joints; the actual face sizes of the blocks are 390 × 190 and 440 × 215 mm respectively.

Clay blocks are made from the same basic material as bricks but to keep them light they are made hollow (see Fig. 4.1). They are available in three thicknesses, 100 mm and 75 mm load-bearing blocks and 62.5 mm non-load-bearing partition blocks, and one face size of 290 × 215 mm actual, which gives a format size of 300 × 225 mm.

All blocks can be supplied with smooth or keyed faces, one or both sides. The keying of the face provides a better background for plaster.

Block partitions are constructed by laying the blocks to stretcher bond in cement mortar which is gauged with lime or proprietary additives to give a joint strength similar to the block strength. Secure attachment of the edges of the partition to the surrounding structure is important in these thin walls and is achieved by block bonding, as described for brick partitions, trying the partition to the external wall with strips of expanded metal laid in the joints or by the use of a metal channel fixed to the external wall (see Fig. 4.1). The top of the partition must also be wedged or pinned to the floor or roof structure. The thickness of block is related to the size of the panel and the table in Fig. 4.1 shows the maximum vertical or horizontal panel dimension for a range of thicknesses of lightweight concrete block partition, assuming adequate edge restraint.

Both lightweight concrete and clay block partitions can be built off a suspended timber floor, which offers great advantages in house design by removing the necessity of positioning the upper-floor partitions over the ground-floor partitions. To take the extra load of the partition, the floor must be strengthened by placing double joists under the line of the partition, and it should be built off a timber plate laid across or along the joists.

Partitions can be built of blocks easier and quicker than of bricks, mainly because of the larger size of the units, coupled with their light weight, but also because they are easy to cut. This saving in labour makes the construction less expensive and is probably the main reason the method is selected.

Another advantage of lightweight concrete blocks is that they afford an easy means of fixing; nails and screws are readily driven directly into the material of the block and give quite high safe working loads.

Clay blocks, however, can present problems in this respect

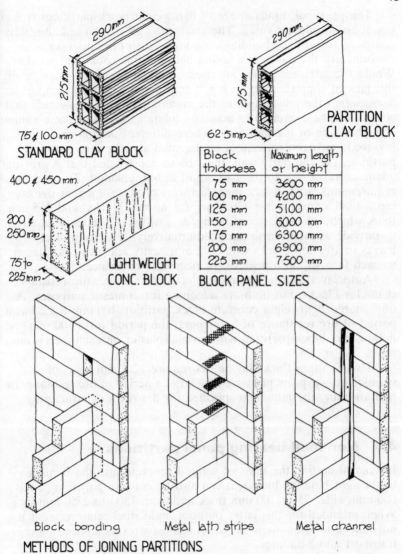

STANDARD CLAY BLOCK

290 mm
215 mm
75 & 100 mm

PARTITION CLAY BLOCK

290 mm
215 mm
62.5 mm

LIGHTWEIGHT CONC. BLOCK

400 & 450 mm.
200 & 250 mm
75 to 225 mm

BLOCK PANEL SIZES

Block thickness	Maximum length or height
75 mm	3600 mm
100 mm	4200 mm
125 mm	5100 mm
150 mm	6000 mm
175 mm	6300 mm
200 mm	6900 mm
225 mm	7500 mm

Block bonding   Metal lath strips   Metal channel

METHODS OF JOINING PARTITIONS

Fig. 4.1 Block partitioning

because of their hollow formation. The only satisfactory way of fixing is by means of purpose-make cavity fixing or toggle bolts. Neither thin lightweight concrete blocks nor clay blocks can support very large loads fixed to their face and, therefore, careful consideration must be given to such items as water tanks, cisterns, sinks and seats on cantilever brackets, and it is preferred that lavatory basins should be fitted with a pedestal.

The principal disadvantage of lightweight block partitions is their low sound-insulating value. The small mass of the wall, deliberately contrived to produce building economies, also presents very little resistance to the passage of sound between one room and another. Where the adjacent rooms are used by members of the same family, this lack of sound insulation is not considered sufficiently important to outweigh the advantages of the material, since it is assumed that control can be exercised to maintain tolerable levels of noise volume on either side of the partition. Where different occupancies are involved, such control cannot be assumed and the intervening partition, i.e. the party wall, must be so constructed that it provides adequate resistance to the passage of airborne sound. To meet these requirements, some block manufacturers produce a higher density block (880 kg/m^3 against 730 kg/m^3 for normal blocks) and 215 mm thick which, when plastered both sides, will give the mass required to provide adequate resistance. Alternatively, a cavity wall of two leaves of 100 mm blocks with a 75 mm cavity and two-coat plaster to each face will also achieve the necessary resistance.

Autoclaved aerated concrete blocks offer longer fire resistance than clay blocks, but both are adequate for domestic purposes. An unplastered lightweight concrete block partition 100 mm thick has a period of fire resistance of four hours; the period for a 100 mm hollow clay block partition with normal plaster on each face is one hour.

The 62.5 mm thick clay blocks require 12.5 mm coats of vermiculite – gypsum plaster – to achieve a period of fire resistance of half an hour, the minimum specified for fire-resisting structures.

4.4 Non-load-bearing panel partitions

Instead of cutting the mass of aerated concrete into the familiar lightweight building blocks, it can also be cut into partition panels 600 mm wide, 75 or 100 mm thick and from 1500 to 2750 mm long. When intended for this latter purpose, mild steel reinforcement is introduced to strengthen the panel against stresses induced during transport and handling.

The panels are erected on end and the vertical edges are butt-jointed using a special glue mortar. The erection procedure is first to apply the mortar, then each panel is placed in position pushed up against a foamed plastic strip in a channel on the ceiling and secured with wedges at the bottom. Finally, the spaces between the wedges are packed with a dry mortar mix and the partition is ready to receive its finish of direct decoration or tiling, or a skin of thin-coat plaster (see Fig. 4.2).

Partitions of lightweight concrete panels impose the same load on the floor as a partition of lightweight blocks and, therefore, the

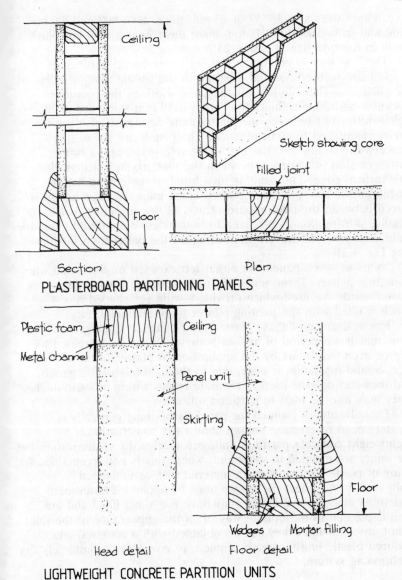

Sketch showing core

Filled joint

Section — Plan

PLASTERBOARD PARTITIONING PANELS

Plastic foam

Metal channel

Ceiling

Panel unit

Skirting

Floor

Wedges — Mortar filling

Head detail — Floor detail

LIGHTWEIGHT CONCRETE PARTITION UNITS

Fig. 4.2 Panel partitioning

floor structure must be suitably strengthened to take it. This additional strengthening is not required for plasterboard partition units.

Plasterboard partition panels consist of two sheets of either 9.00 mm or 12.7 mm plasterboard separated by a fibrous cellular core. Panel sizes are: 38, 57 or 63 mm thick, 600, 900 or 1200 mm wide and from 1800 to 3048 mm high; there is also a 50 mm thick panel in the one size of 900 × 2350 mm.

The first step in erecting a cellular plasterboard partition is to fix a sole plate to the floor and a batten to the ceiling along the line of the partition. The sole plate is the same width as the partition thickness and 50 mm thick; the batten is 19 mm thick and of the right width to fit between the plasterboard facings; it also has the angles chamfered to assist erection. The panels are cut to fit between the ceiling and the top of the sole plate, and a timber batten or stud is tapped in between the plasterboard facings along one vertical edge. The panel is then fitted to the ceiling batten, placed on the sole plate and pushed along until the edge batten is forced between the plasterboard facings of the adjoining panel. Finally, skirting boards are fixed to both sides, by nailing to the sole plate, and serve to prevent the bottom of the panel from moving (see Fig. 4.2).

Joints between panels are either left exposed or concealed with a jointing plaster. If the joints are to be hidden, tapered-edged plasterboards are used which provide a wide vee-shaped recess which is filled with the jointing plaster to give a flat surface.

Fire resistance of these plasterboard panels is quite good. The basic unit has a period of half an hour which can be increased to one or even two hours by the application of plaster to the panel face. Sound insulation is poor, due to the lightness of the panels, and they can only be used to separate rooms within a house or flat (they may also be used to partition offices).

The advantages claimed for panel partitioning generally is greater speed of erection and an almost dry construction. Lightweight concrete partition units can receive direct decoration but the joints are inclined to show and, where this is not acceptable, a lining of plaster or heavy sheet material such as embossed wallpaper, p.v.c. sheet or hessian must be applied. Plasterboard panels, as already mentioned, may have the joints filled and are then decorated in the normal way or, if the appearance of the joint is not objectionable, they can be obtained with a coloured and textured plastic finish, already applied, to give a fast, completely dry partitioning system.

4.5 Timber stud partitioning

Studwork is a different approach to the task of dividing the enclosed space in that it is a composite structure comprising the timber framework to provide the support and a facing material applied to the framework to complete the partition. Stud partitions can be used as structural elements, but usually they are considered as non-load-bearing.

The members of the framework are the studs, which are the vertical members, the bottom plate or cill, the top plate or head, which hold the bottom and top ends of the studs respectively, and the noggins, which are fixed between the studs both to stiffen them and to provide additional fixings where required (see Fig. 4.3). The size of the timbers in a non-load-bearing partition is determined by the need for the structure to possess the necessary stiffness within the storey heights. In domestic work, 50 × 100 mm studs are usually provided but, with low ceilings, 50 × 75 mm studs may be adequate. In some situations, 38 mm studs may be used but, if the facing is to be nailed, 50 mm is to be preferred since it provides a greater edge clearance for the nails where the facing sheets meet at the stud. The cill, head and noggins are invariably the same size as the studs. Double studs are fixed up each side of door openings to provide the additional stiffening required at this point to take the force of the swing and the slam of the door.

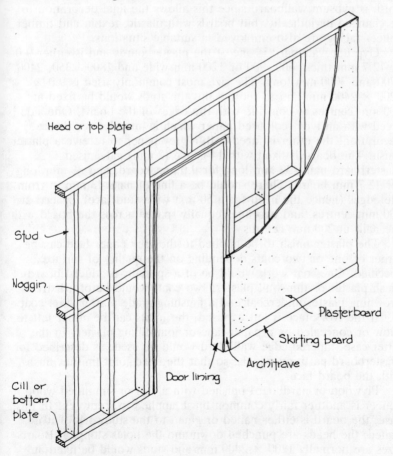

Fig. 4.3 Stud partitioning

The erection procedure is to nail the cill and head to the floor and ceiling, set the studs in position between them, skew nail them into place and finally to fix the lines of noggins at a maximum spacing of 1200 mm. Alternatively, the framework can be cut and assembled first and then erected in its final position. Prefabrication of the studwork has the advantage that it can be made up in advance of the time it is needed and out of the way of other operations, but getting the large panels into place can raise problems and they must be undersized to allow a positioning tolerance, which must then be packed out to obtain a firm fixing. Noggins may be fixed in a herringbone fashion, but usually they are fixed level and staggered to allow the studs to be through nailed to them.

Any of the great variety of sheet materials may be used to line the studwork. Selection is purely a matter of preference based on the finished appearance desired. Possibly the most frequently used finish is gypsum wallboard since this allows the final decoration to be changed periodically, but boards with plastic, textile and timber veneer facing are also employed in suitable situations.

Gypsum wallboard is one of the plasterboards and is either 9.0 or 12.7 mm thick, 600, 900 or 1200 mm wide and 1800, 2350, 2400, 2700 and 3000 mm long. The size most commonly used is 9.0 × 1200 × 2400 mm and to receive this the studs would be fixed at 400 mm centres to coincide with the edges of the board. One side is faced with an ivory-coloured paper which is intended for direct decoration, the other is faced with a grey paper to receive a plaster finish. The board is fixed with 14-gauge (2 mm) galvanised plasterboard nails, 32 mm long for 9.0 mm board, and 38 mm long for 12.7 mm board. There should be a line of nails 13 mm in from the edges (hence the need for a 50 mm wide stud face), spaced at 100 mm centres, and two lines equally spaced across the board with the nails at 200 mm centres.

The plaster finish to be applied to the grey paper face can be either of one or two coats depending on the quality of surface specified. One-coat work would be of a specially-produced board-finish plaster or thin-wall plaster; two coats would be of a haired browning plaster undercoat and a finishing-grade plaster skim coat.

If the ivory face is left exposed, the joints can be either left to show or concealed by a paper-tape or joint-filling plaster. In the latter case, tapered-edge wallboard would be used, as described for plasterboard partition panels, so that the filled joint finishes flush with the board face.

Plywood or hardboard finished with a real or simulated wood veneer is another fairly common finish applied to studwork. In this case, the board is either nailed or glued to the studwork; if it is nailed, the heads are punched down and the holes stopped. Board sizes are normally 1200 × 2400 mm and studs would be fixed at

400 mm centres. Usually the boards are supplied completely finished and no further work is required, but care must be taken in handling and storing them to avoid damaging the pre-finished face.

The advantage of timber stud partitions is that they are light, easily and quickly constructed and, because the material is readily worked, they can be fitted or built to any shape given or required. With the right facings they can achieve a satisfactory level of fire resistance, and a double system of studwork with at least 200 mm spacing between, plus a double layer of 19 mm gypsum plasterboard to the room face of each set of studs, can be shown to have sufficient sound insulation for use as a party wall. Without this rather excessive construction for sound insulation, studwork is not very good in this respect, as the light structure and thin facings vibrate quite readily and offer very little resistance to the passage of sound. Another disadvantage frequently encountered by owners of houses with timber stud construction is fixing objects to the wall. If the screw or nail can be positioned to coincide with a stud or noggin, a secure fixing can be achieved but, since the area of the stud faces is only about one-eighth that of the wall, the likelihood is that the fixing point misses the timber framing and a cavity-fixing method must be adopted, which puts all the load on the facing material. In consequence, the load which can be placed on these fixings is restricted and domestic equipment such as bookshelves, cupboards or the pivoting bracket supports for television sets may prove to be in excess of what the facing material can support.

Being constructed of the natural organic material – wood – timber stud partitions are prone to movement, particularly in the early days of the building's life, when the whole structure is settling down to a temperature and moisture equilibrium. This possible movement must be taken into account when lining studwork with any rigid or brittle finishes. It can, for instance, lead to cracks developing in plastered surfaces, particularly if a thin skin coat only is used over wallboard. It also affects the application of wall tiling in that both the adhesive and the grout must be carefully selected to accommodate a certain amount of relative movement without loss of adhesion. If the partition is tiled, because water will be directed at it (as in a shower enclosure) even greater care is needed because the most minor failure of the continuity of tile and grout could allow water to penetrate, leading to greater movement of the structure, which will aggravate the failure of the tiles to exclude the water and, eventually, to decay of the timber.

Chapter 5

The industrialisation of domestic construction

5.1 Factory production of building components

Progress in the design and capacity of machinery, greatly improved methods of transport and a more rational approach to the design of buildings have combined to produce considerable changes in the production of buildings. Today's sophisticated equipment can now produce the large components required for building purposes quickly and economically and, furthermore, if it is in a factory environment, production is continuous no matter what the weather is like on the site. Without a corresponding sophistication in the methods of handling and transportation, the progress which has occurred in industrial production would not have been possible. A controlling factor in the design of any prefabricated component is the ease with which it can be picked up, carried and placed in its correct position. Neither the developments in fabrication nor the improvements in transportation would have occurred without there being a demand for industrialised components. This demand has arisen through building design based on construction techniques developed from the continuous search for more efficient and economic building methods rather than on the dictates of architectural style.

Timber, concrete or steel all lend themselves to the production of building components in a factory but, for domestic work, the majority of prefabrication is carried out in timber. Timber is light for its strength and easily handled and fitted, thus it is most suitable for the small scale of domestic work, which is well within the economic limits of structural timber units.

A further boost to the factory production of timber units has been the introduction of stress-graded timber which, in conjunction with developments in methods of jointing, has allowed a reduction in the size of the timber members needed to carry a specific load and, consequently, a reduction in the weight of a component.

5.2 Stress grading of timber

Structural softwood is now graded in the UK and many other countries according to the stress-carrying capacity of each individual piece. Such stress grading has had the effect of raising the degree of reliability which can be placed on the material. The design of structural members in ungraded timber must be based on the load-carrying capacity of the weakest member which may be found in the batch of timber supplied. Inevitably, this meant that the majority of the timber used was greatly understressed and consequently oversized, but nobody knew which one was going to be the critical piece. Stress grading now eliminates the very weak timbers and, therefore, the allowable safe working loads can be increased to nearer the theoretical maximum for the material.

For structural purposes, the prefect timber is straight-grained, uniform in density and free from any knots or defects but, as is clear to any observer, trees do not grow dead straight, nor do they grow evenly and they do have branches (which is what produces the knots). A clear perfect specimen of softwood is not very common and, therefore, timber is graded within permitted degrees of slope of grain and size and spacing of knots. It should be noted that the actual species of tree, other than being deciduous, is not the concern of stress grading: all that is determined is the load-carrying capacity of the individual piece of timber.

There are two sets of grades applicable in the UK and a set of rules for grading timber in Canada. Appendix A of CP 112 (1971) defines a series of numbered grades of timber (40, 50, 65, 75 and a composite 40/50 grade) but these are being phased out in favour of the grades in BS 4978 : 1973 'Timber grades for structural use'. The British Standard defines two principal grades for solid timber: general structural (GS) and special structural (SS).

The grade requirement can be met by visual grading or a machine can be employed to test each piece, in which case the grades are designated as machine general structural (MGS) or machine special structural (MSS). To these the BS adds two further machine grades, M50 and M75.

The Canadian grading rules are administered by the National Lumber Grades Authority (NLGA) and approved by the Canadian Timber Standards Administrative Board. Currently, there are three grades imported into the UK and these are further subdivided as shown below:

Grade	Subdivision
Structural joists and planks grade	Select Structural No. 1, No. 2 and No. 3
Light framing grade	Construction, Standard or Utility
Stud grade	No subdivision

Each graded piece of timber must be marked at least once within its length. Visually-graded timber. has a mark to identify the grader and the company responsible for the grading. Machine-graded pieces must show the licence number of the stress-grading machine, the strength class under BS 5268 assigned to the piece, the species and grade, the BSI kitemark and the reference–BS 4978.

Visual grading is based on the size and spacing of knots. The presence of knots in timber has the effect of weakening it because they interrupt the line of the grain along which the stress flows. The larger and closer the knots, the greater the weakening effect.

Grading is based on the knot area ratio (KAR) and the spacing or longitudinal separation. The KAR is the ratio between the cross-sectional area of the knot and that of the timber and is more critical if the knot is in the upper or lower quarter of the cross-section than if it is in the middle. This disposition of the knots is referred to as the 'margin condition'.

For a timber to be graded SS, the KAR must be less than one to five and for GS grade the KAR is beween one to five and one to two depending on the margin condition. Where two knots or groups of knots occur in which the KAR is over 90 per cent of the permitted maximum, their longitudinal separation must be not less than half the timber thickness for SS grade and not less than half the timber width for GS grade.

Machine grading is based on the stiffness of the timber and is tested by either measuring the amount of deflection when a standard force is applied or measuring the amount of force needed to produce a standard deflection.

BS 5268 : Part 2: 1984 groups these softwood stress gradings and the Canadian gradings into five strength classes, numbered SC1 to SC5, to each of which it assigns a design value for the grade stress and the modulus of elasticity applicable.

5.3 Jointing methods

The traditional method of joining two timbers together was to cut and shape each so that the one fitted into the other, to be retained by a peg driven into a hole. Typical of such a joint is the mortice and tenon in which a rectangular hole, or mortice, is cut in one member and a tongue, or tenon, cut on the end of the other to fit into the mortice. After assembly, a hole is drilled through the timber each side of the mortice and through the tenon, and the peg, or dowel, driven in. A direct result of this jointing method is a considerable reduction in the cross-section of area of the timber members, due to the removal of material to make the mortice and the dowel hole. To compensate for which the whole section must be increased in size.

These jointing techniques are still widely used in the production of joinery units, but structural members, of critical cross-sectional area, are generally overlapped or butted and joined with nails, screws, bolts, bolts and timber connectors or truss or gangnail plates (see Fig. 5.1).

Nails and screws separate the grain without removing any of the timber and, therefore, have less of a weakening effect. If the nail or screw is driven too near to the end of a member, the grain separation becomes a split and the fixing loses a lot of its strength. Screws achieve a stronger grip than nails but are more expensive to buy and to fix.

Bolts on their own are very inefficient because the whole strength of the joint relies on the shear resistance of the timber parallel to the grain, and this is usually low in most species (see Fig. 5.1). The area of timber subjected to shear forces at a joint can be greatly increased by the use of either toothed plate or split ring connectors fitted round the bolt. The toothed plate connector consists of a galvanised steel plate, drilled to fit the bolt and cut into points which are bent alternately up and down around the edges to form teeth. This is fitted on to the bolt between the timber faces and the teeth bite into each member as the bolt is tightened. The split ring connector is a hoop of steel with a small gap at one point. It is fitted into a groove specially cut round the bolt hole on each of the matching timber faces; the small gap or split allows the ring to be slightly compressed into the groove, ensuring a tight fit. When the ring is in place, a bolt is passed through the central hole and tightened to complete the joint. Split ring connectors produce a greater increase in shear resistance at the joint than shear plate connectors but, as they require more work to fit, are more expensive.

Truss or gangnail plates are galvanised steel plates in which nail shapes have been punched and bent up. They are used to connect the members of a trussed timber frame and are more suited to industrialised production than to use on site. To make the joint the timbers are cut and butted together and then a truss plate is hydraulically pressed into each face to span the joint line and to connect the timbers. The whole process is now highly automated and the final truss accurately assembled in a large jig. Not only is this method of connection very economic in factory production, the design and assembly of the trussed frame is greatly simplified by all

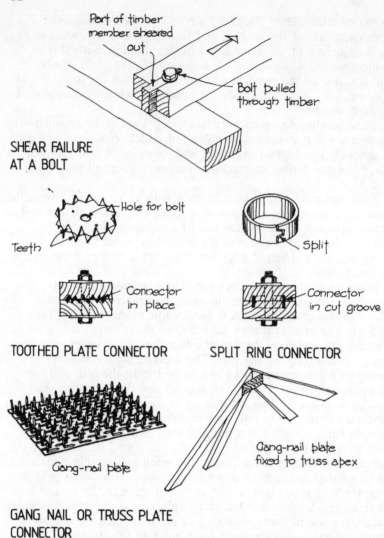

Fig. 5.1 Timber connectors

the timbers being in the same plane. Where timbers are overlapped, problems are created by there being differing numbers of overlapping members at various points. This makes it necessary to introduce short timbers, between the main members, as packing to keep them all parallel to each other, thus ensuring that the overlapping faces are fully in contact.

5.4 Roof frames

The most noticeable effect of industrialisation on house building has been on the construction of roofs. Here, the advantages of stress-graded timber, gangnail plates and factory production have combined to give a building technique which has almost completely replaced the traditional methods.

In the past, the roof covering was supported at the appropriate pitch by rafters spaced at 400 to 450 mm centres. These rafters were supported by purlins running along the roof slopes below the rafters. The purlins were supported by trusses in the same plane as the rafters, at 1500 to 2000 mm centres, bearing on to the walls (see Fig. 5.2). This method has now been replaced by the trussed rafter system where each rafter is part of a trussed frame and no purlins are required. For house roofs, these trussed rafters can easily be transported by road and, therefore, a complete roof structure can be prefabricated off-site.

The most common form of prefabricated trussed rafter is the Fink Truss (see Fig. 5.2), and these are usually delivered to the site in banded lots, a complete roof in one lorry load. Where a crane is available, either the trusses are lifted to roof level in one bundle or the roof is completely assembled at ground level and then lifted on to the walls.

The first operation is assembling a roof frame in place is to position an end truss on the timber wall plates bedded on top of the wall. This truss is set vertically, nailed to the plates and temporarily supported. It is followed by another three or four trusses, each accurately located on setting-out marks on the wall plates. Once these trusses are in place, they are held at the correct spacing by a temporary batten along the top temporary diagnonal bracing to keep them upright. The rest of the trusses are then fixed and retained by a temporary batten at the top. With all the trusses in position, permanent diagonal bracing is fixed to the underside of the rafters, running from the apex of the end trusses to the feet of the fourth or fifth truss along. The temporary bracing and battens are then taken off, the roof felted and the tile battens fixed to complete the structure. To straighten the bottom tie member of the truss for convenience in fixing the ceiling board, a timber binder is run along the length of the roof and nailed to each truss.

A notable difference between a traditional roof and a trussed rafter roof is that the latter has no ridge board. This member has no structural value and was provided for the convenience of the carpenter in nailing the individual rafters together. With the prefabricated frame, the joints are already made and the ridge board can be dispensed with. If one goes further back into history, to medieval days, when nails were not so readily available, one finds that the roofs then did not have a ridge board–the carpenters

54

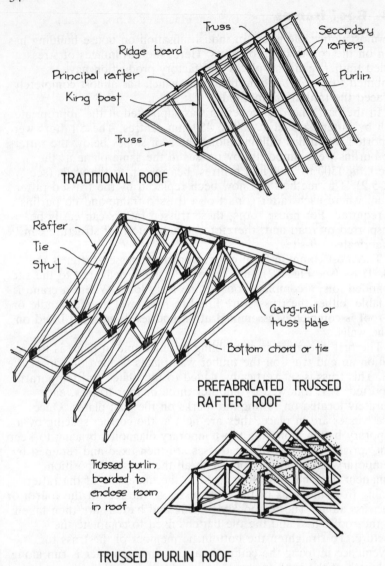

Truss

Ridge board

Principal rafter

King post

Truss

Secondary rafters

Purlin

TRADITIONAL ROOF

Rafter

Tie

Strut

Gang-nail or truss plate

Bottom chord or tie

PREFABRICATED TRUSSED RAFTER ROOF

Trussed purlin boarded to enclose room in roof

TRUSSED PURLIN ROOF

Fig. 5.2 Roof frames

notched the rafters together and secured them with a peg.

Industrialised frames, using stress-graded timber and gangnail plates, are highly efficient in the use of material. Each member is carefully sized to carry its predicted stress with the maximum economy. For this reason, none of the timbers should be cut, notched or mutilated in any way, nor should additional loads be placed on the frame, unless the manufacturer is first consulted. Cutting of the trusses can be occasioned by the need to provide openings for trap doors, chimneys, roof lights etc., and additional loads can be imposed by water tanks and other equipment being placed on the bottom tie member.

An alternative approach to roof design is the trussed purlin structure (see Fig. 5.2). In this system, the rafters, as in the traditional roof form, are supported by a purlin but, instead of the purlin consisting of a single piece of timber which required supporting at frequent intervals, it is framed up into a girder formation which can span the length of the roof from gable wall to gable wall. Trussed purlin frames can be threaded through prefabricated rafter frames or they can be used in conjunction with on-site assembled loose rafters and ceiling joists. An attraction of the latter structure is that the centre of the roof void is left uncluttered by roof timbers can be used for storage or accommodation, provided that these loadings are anticipated in the design.

5.5 The use of timber-based sheet material

A product which was not available to builders and designers in the past is the timber-based board or sheet. Modern technology can now reduce a complete log to a thin veneer and then glue these together to produce plywood or blockboard; or the timber can be reduced to particles or flour, mixed with resins and pressed into sheets to give hardboard or chipboard. In all forms, the finished panel is larger than would have been possible to produce from the natural timber, and much more stable.

The introduction of these sheet materials has made possible forms of construction which could never have been achieved in the past.

From the structural point of view, the predominantly useful feature of these sheet materials is their rigidity in the plane of the board, sometimes known as the 'diaphragm effect'. It can be appreciated that, while it is easy to bend a sheet of plywood by laying it flat across two supports and placing a load in the middle, it is much more difficult to produce any distortion if the sheet is stood on edge. In this attitude, it only fails by the sheet twisting or buckling and, thereby, departing from the original vertical position.

Structural designers make use of this characteristic in timber beams, by combining plywood strips on edge with solid timbers along the top and bottom, to restrict any tendency to buckle. The forms these take are 'I' section beams (in one version the plywood is corrugated for additional stiffness) and box beams (see Fig. 5.3).

Another structural technique which is applied to the use of sheet

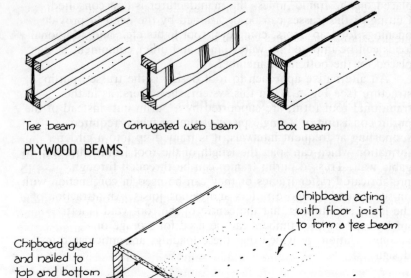

Tee beam   Corrugated web beam   Box beam

PLYWOOD BEAMS

Chipboard acting
with floor joist
to form a tee beam

Chipboard glued
and nailed to
top and bottom
of floor joists

**STRESSED SKIN
FLOOR PANEL**

Fig. 5.3 Prefabricated timber floor members

material is stressed skin construction. The principle of stressed skin
design is that if a thin membrane is stretched across and rigidly
secured to the top and bottom edges of a series of ribs, the resulting
complex is stronger than the collective strength of the ribs acting on
their own. It is a principle most clearly demonstrated in the balsa
wood and tissue paper construction of model aircraft, which achieves
a strong structure from weak materials. This principle has been
applied to the manufacture of prefabricated floor panels in which a
large part of, or the complete upper floor of, a building is made up
as a transportable unit. In this case the top and bottom edges of the
floor joists are planed to ensure that they all line up perfectly and
to provide a smooth surface. These regularised floor joists are then
spaced at 400 mm centres and glued and pinned between sheets of
chipboard. When complete, the chipboard and the joists act together
to form a series of tee beams with wide top and bottom flanges of
chipboard and webs of softwood.

The upper layer of board provides the floor deck and the lower layer becomes the ceiling surface for the room below. By this means, significant savings can be made in the quantity of timber required to span any given distance.

5.6 Prefabricated infill panels

The economic advantages of industrialisation are only fully realised when the component is repeated a sufficiently large number of times to offset the cost of setting up the machinery involved. Prefabricated components, such as trussed rafters, have a universal application to bungalows, houses, flats and many other buildings and, consequently, repeat orders, with only minor variations, are readily obtainable. Prefabricated infill panels are much more closely related to the design of both the plan arrangement and the elevational treatment of the building. Therefore, they rely on there being a large number of the same building units required to achieve the repetition necessary to produce maximum cost efficiency. For this reason, the use of infill panels is mainly restricted to buildings with multiple occupancy such as blocks of flats or maisonettes.

When designing a flat, the architect must bear in mind that there will be a similar flat each side of the one he is designing and probably one above or below or both. This creates two design factors: firstly, the windows can only be located in the front and back walls and, secondly, the side walls and floors must be imperforate, sound insulating and fire resistant. The logical outcome of the effect of these factors is a building with solid sound-insulating and fire-resisting walls running from front to back which, because of their solidity, are strong enough to support concrete floors spanning from wall to wall. This provides the load-bearing structure and all that is then required are light end walls to complete the enclosure and internal partitions, to subdivide the space into rooms. This system is known as cross-wall construction (see Fig. 5.4).

It is in the manufacture of the repetitive end walls or infill panels that industrialisation can offer advantages. These can be made of concrete slabs or metal or timber framing with a variety of claddings. Timber, by reason of its lightness and ease of working, is a popular choice for infill panels. The cladding can also be of timber, in boards or, alternatively, metal lath and render, plastics or metal sheets or panels, or tile hanging can be used. Figure 5.4 shows details of a typical timber-framed infill panel incorporating standard timber windows and clad with tile hanging.

The performance requirements that must be considered when detailing an infill panel concern both the way the finished product, in its location, satisfies the needs of the building occupant, and the demands of the methods used for prefabricating, transporting,

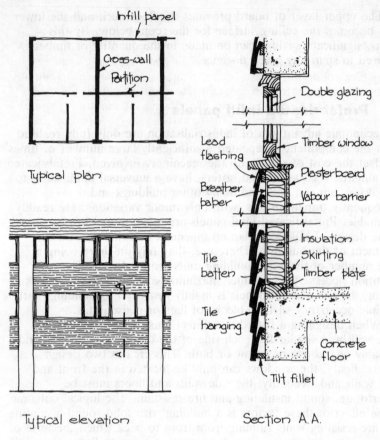

Infill panel

Cross-wall

Partition

Typical plan

Double glazing

Timber window

Lead flashing

Plasterboard

Breather paper

Vapour barrier

Insulation

Tile batten

Skirting

Timber plate

Tile hanging

Concrete floor

Tilt fillet

Typical elevation

Section A.A.

Fig. 5.4 Timber infill panel

handling and locating the panel. The problems of the manufacturer and the builder vary considerably depending on the nature and size of the contract, the location of the site in relation to the factory workshop, and the size and organisation of the firms involved. Domestic requirements are more clearly defined. The Building Regulations impose standards of fire resistance, and sound and thermal insulation. The occupant and the owner would expect an adequate resistance to the effects of weather, the admission of adequate natural light and freedom from draughts.

Fire-resistance requirements in the situation in which infill panel walls are found are not usually difficult to meet, since the distance between the front or back walls of a block of flats and the site boundaries is usually large enough to prevent any serious hazard to adjacent properties in the event of a fire.

Sound-insulation requirements are concerned with noise from

one flat passing out through the infill panel, round the end of the sound-insulating crosswall and in through the infill panel of the adjoining flat.

It is not easy to make a lightweight panel resistant to the passage of sound and if the occupant opens a window, even slightly, all the designer's efforts to contain the sound would be nullified. The solution to this problem lies in reducing the amount of sound which can leak round the end of the cross wall. This is done by either projecting the cross wall at least 460 mm beyond the face of the infill panel or by providing return walls on the end of the cross wall to give a minimum separation between the panel walls of 690 mm. This is the solution shown in Fig. 5.4 and has the added advantage that the return walls stiffen the cross wall.

Thermal insulation is achieved by the introduction of an insulating quilt between the members of the timber frame, and can be further improved by double glazing the windows and fitting draught-excluding weatherstrips to the opening lights (see Fig. 2.2). It should be noted that the vapour barrier, shown in Fig. 5.4, is very important in a highly insulated timber framework. Its purpose is to prevent moisture vapour from penetrating to the outer, cooler parts of the structure where it may condense into moisture droplets which would soak into the insulation, thereby reducing its value and also creating a high risk of decay occurring in the timber. For this purpose the vapour barrier must always be on the warm side of the insulation.

Resistance to weather is a factor controlling the choice of external cladding. As an example, the tile hanging shown in the illustration will satisfactorily withstand the effects of rain, snow, sun and frost but may not be suitable on very exposed sites, where the negative wind pressure on the leeward side of the building could lift the tiles off.

Chapter 6

Industrialisation of joinery and services

6.1 Factory production of joinery

A distinction is made in the building industry between the timber trades of carpentry and joinery. A carpenter works with large size timbers, often unplaned, and connects them with nails or bolts – his area of concern is usually the structure of the building. A joiner's work involves much finer detail: his timbers are usually small and planed or moulded, he is skilled in the making of timber joints (hence his name) and is generally concerned with windows, doors, stairs, cupboards and carefully-finished fittings of all kinds.

Great accuracy is needed to make a satisfactory timber joint – both halves must be cut exactly to line and square, so that the faces meet with precision. The degree of accuracy required can be easily achieved and continuously maintained by woodworking machinery, but it takes much longer to set up a machine than to pick up a hand tool. Because of this, the special orders, for one or two components, are made by hand, with the assistance of the general woodworking machines such as the various saws, planer, thicknesser and morticer. For larger orders involving long runs, more sophisticated equipment is employed in which many more operations are carried out.

From this, it can be seen that the production of joinery is a process which relies heavily on the use of both static and portable machinery. As such, it is carried out in a workshop (on some contracts this workshop may be set up on the site) and naturally lends itself to industrialisation. All that is required is the standardisation of designs and sizes for the full advantages to be gained. Not all joinery components can be standardised — most buildings require the joiner to get out his hand saw and his chisel and make up a few special items – but many of the carefully-finished timber components serve identical functions in all houses and can, therefore, be produced as standard or as a range of standards.

Typical of industrialised joinery elements are windows, stairs, door sets and cupboards. Seldom are windows specially made for housing purposes now, the variety available within the standard ranges which have been produced for many years usually satisfying the designer's requirements. Although the manufacture of staircases has made use of machinery for a long time, the introduction of 2600 mm as a preferred floor-to-floor dimension in housing has encouraged the joinery manufacturers to offer standard stair flights. Doors, like windows, have been factory-made for many years but a relatively recent innovation is the pre-hung door set. The cupboards in a house are found mainly in the kitchen or the bedroom. Improvements in the design of kitchen fittings and the provision of built-in bedroom furniture has been stimulated by the ready availability of attractive units as a direct result of the industrialisation of their production. All these components are dealt with in more detail in the following sections.

6.2 Windows

Basically, there are two parts to a window, the fixed frame which is built into the wall, and the opening lights which are attached to the frame. The method of attachment determines the way the window opens and can be hinged, pivoted or sliding. Whatever the type of size of window, the range of sectional profiles of timber needed can, by careful design, be reduced to half a dozen or so. The situation of a large number of windows being produced from a small number of timber sections lends itself to industrialised production in machines which accept sawn timber at one end and eject planed and moulded lengths of window section at the other. These are then cut and jointed in other machines, leaving only the actual final assembly to be done by hand.

Manufacturers now offer several ranges of window in differing prices depending on the complexity of the frame and the material from which it is made. The cheapest are the standard windows of softwood and the most expensive are the high-performance windows in hardwood. Standard windows are based on designs developed by the English Joinery Manufacturers Association (EJMA) and are covered by BS 644 : Part 1 : 1951. They are for single glazing, and the overall sizes of the frame are based on a 300 mm module. Most

62

manufacturers modify the standards recommended, following the basic principles, but improving the section detail and adjusting the overall sizes slightly (see Fig. 6.1). The high-performance joinery details have been developed to meet increasing standards of weather exclusion. The main differences between standard and high-performance frames are that the latter are made to receive double glazing sealed units and incorporate a draught-excluding weatherseal. The frame sizes mostly follow the same 300 mm module as the standard frames.

The over-riding advantage in the use of standard windows is cost. The level of investment in sophisticated equipment by manufacturers is very high and there is considerable expense in setting all the cutters, guides, stops etc. to produce the joinery but,

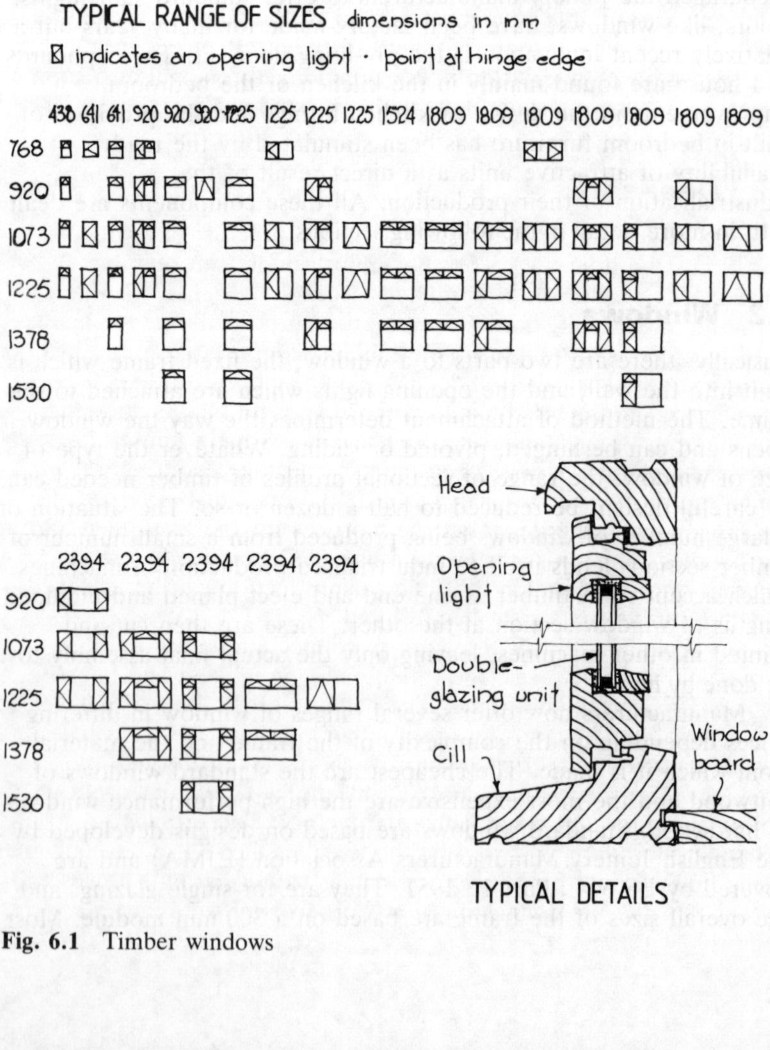

TYPICAL RANGE OF SIZES dimensions in mm

⊠ indicates an opening light · point at hinge edge

Head
Opening light
Double-glazing unit
Cill
Window board

TYPICAL DETAILS

Fig. 6.1 Timber windows

because the finished window is produced in large quantities, it still costs less than the same frame would do if it was made by hand. A further economy results from a saving of time. Ranges of windows are held in stock by merchants and, generally, a builder can go and collect his requirements for a single house as needed. Large quantities may have to be ordered but still it would only be a matter of days before the builder received them. If he had to make them to order, starting when the contract was signed, delays would be likely as the time taken to make the window frames could well exceed that required to lay the foundations and raise the walls to cill level.

There is a slight advantage in the process of glazing if the windows are standard in that the glass size is also standardised. As a result, it is not necessary for the glazier to go round the house and measure for each pane of glass separately. All he needs to do is to order the glass by the reference number of the window and the glass merchant will cut the required panes to the sizes set out in a reference table. Some of the window manufacturers will also supply the glass or double-glazing units with the window.

6.3 Door sets

Framing up a traditional panelled door is a process, like windows, which must be done in a workshop. Flush doors can only be satisfactorily produced on machinery specially designed for the purpose. As a result, doors for domestic purposes have, for many years, always been bought by the builder from a large joinery works rather than made in the builder's own workshop. The door frames or linings may also be obtained complete or the builder may just buy the appropriate timber sections and make them himself. After the frames or linings have been built in and the plastering is complete, the joiner will arrive with the doors, cut the rebates for the hinges, hang the door, cut the mortices for the lock, fit it and fit the handles. This on-site operation has been examined to see whether improvements in efficiency of production can be achieved, and the pre-hung door set has resulted.

As explained above, the work in connection with a door is carried out in two stages: the frame or lining is fixed and later the door is attached to the frame. The hanging of the door is delayed to avoid possible damage from the weather (because the roof may not be on) and from materials or barrows being taken through the opening. With a pre-hung door set, all the work is carried out before the door and frame leave the factory. The problem of protecting the door is solved by the use of either loose-pin hinges or rising butts. Loose-pin hinges are similar to normal butt hinges except that the pin connecting the two halves of the hinge is removable.

Rising butts are hinges in which the bearing face between the two halves of the hinge is set at an angle so that as the hinge is opened it also rises. These are fitted to fire doors and the weight of the door, pushing the leaf of the hinge downwards, causes it to travel round the sloping bearing face, thus closing the door. It is necessary that fire-resisting doors are self-closing for them to be effective. With either type of hinge, the door can be taken off the frame and stored in a place of safety until later in the construction programme, when the only work required is to either line up the two halves of the hinge and replace the loose pin or place the door leaf of the rising butt on to the pivot of the frame leaf.

The simplicity of this operation becomes even more attractive if pre-finished doors are specified. Many joinery works now produce hardwood veneered and polished doors in a finished state, and if they are pre-hung they can be left off until almost the last few days of the contract, to avoid all possibility of damage by anybody.

6.4 Stairs

The industrialisation of staircases for houses was given an impetus, as already mentioned, by the adoption of the preferred floor-to-floor height of 2600 mm. This was first introduced by the Ministry of Housing and Local Government (as it then was) at the same time as the move towards metrication started, in the early 1970s. It was part of a complete package of metric dimensional co-ordination for local authority housing. Up to this time, staircases had to be made to suit whatever the floor-to-floor height happened to be. Obviously, the staircase must be an accurate fit between floors, otherwise, if it rises too far, there is a small step down to the first floor, which is a danger to someone about to descend and, if it does not rise far enough, the flight is lifted which brings the top level, but creates a higher step at the bottom, over which people will trip.

The manufacture of a staircase has always been a workshop or factory job, and gradually machines were introduced to make the task easier, faster and less expensive. The joiner would carefully measure the floor-to-floor height and then set out the treads and risers required to suit the overall rise. These were drawn out on the strings – the two timbers which enclose the sides of the flight, and into which the treads and risers are fitted (see Fig. 6.2) – and cut out to house the ends of the treads and risers. The flight was then glued, assembled, wedged and screwed ready for transport to the site.

Standardisation has not changed the structural design of a staircase flight, the treads and risers are still housed into strings and wedged, but by reducing the variety of total rises to one and the number of steps to thirteen (thirteen risers but only twelve treads,

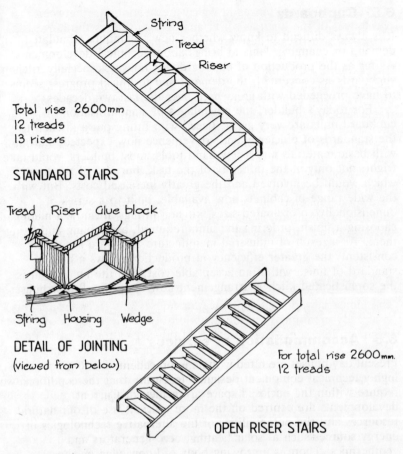

Total rise 2600mm
12 treads
13 risers

STANDARD STAIRS

Tread Riser Glue block

String Housing Wedge

DETAIL OF JOINTING
(viewed from below)

For total rise 2600mm.
12 treads

OPEN RISER STAIRS

Fig. 6.2 Timber stairs

because the upper landing becomes the thirteenth tread) the setting-out time has been almost eliminated and flights can be produced in batches rather than one at a time, with consequent improvements in efficiency and reduction in cost.

The advantages to the builder are similar to those relating to standardised windows: immediate availability of the component when required and the avoidance of tying up his workshop and joiners in specially producing an article which is very similar in all houses whether or not they are standardised.

Although the flight is the same in all houses, variations in the balustrading and handrail are possible. Most manufacturers offer a range of newel posts, handrail parts, balustrade spindles and rails and fittings which can be combined to the individual choice of the building designer.

6.5 Cupboards

It is always difficult to know whether new techniques stimulate
demand or changing demand brings about technical developments.
As far as the production of cupboards for houses, especially kitchen
cupboards, is concerned, the demand and technical progress seems
to have proceeded with great pace over the last two decades.

For today's builder, the advantages of standardisation of
cupboard units are very great. Faced with fitting out a kitchen to
the standards of efficiency and convenience now expected, starting
with basic materials and a few hand tools, most builders would take
fright, not only at the difficulty of the task, but also at the time
which would be involved and the greatly increased costs. But with
the wide range of cabinets now available, built to a series of
dimensionally co-ordinated sizes, kitchen fittings are now a matter of
choosing a finish, ordering the units required, assembling and fixing
them. As a result of industrial manufacture, sizes are absolutely
consistent, the greater efficiency in production allows a high
standard of finish within an acceptable cost, and the final result is
the sophisticated kitchen arrangement now found in all new houses.

6.6 Accommodation of services

Present day houses are fitted with an unprecedented amount of
highly technical equipment designed to ensure that the conditions we
require within the enclosed space are achieved. Current
developments are centred on the more efficient use of our natural
resources and the potentialities of the alternative technologies of
energy sources such as solar heating, aero-generators and
geothermics. There is a growing body of knowledge on those
alternative resources and the accommodation of them in future
houses will, inevitably, considerably influence their design and
construction.

Current domestic practice, with regard to the accommodation of
services, tends to treat the equipment and its linking pipes and
cables as 'add-on items', to be attached to the house structure at a
late stage in the building programme. Pipes running across the face
of the wall, radiators occupying useful wall areas, electrical cables
snaking untidily across pitched roof ceiling joists are examples of
this 'add-on' attitude. Even where services are concealed, it tends to
be a make-shift provision such as electrical cables irretrievably
buried in plasterwork, structural floor joists weakened by notches
cut to take water pipes etc.

The most common method of space and water heating provided
in houses at the present time is low-pressure hot-water radiators
using a gas- or oil-fired boiler as the energy source. The efficiency

of the system is very variable and depends on many factors, but it is simple in principle and easy to install in both existing and new houses. Its accommodation requires very little provision to be made in the house, which is why it can be installed in existing properties. The boiler (usually sited in the kitchen because that is where water has always been heated) requires a duct for the disposal of its combustion gases and a supply of combustion air. A chimney for the gases and an air brick in the room containing the boiler will suffice but, if no chimney is available, balanced flue appliances can be used which take in their combustion air and expel their combustion products through a double duct fixed in the wall behind. The water tanks can be accommodated in the roof, but provision must be made to support the weight of water contained; the hot-water cylinder can be placed in a cupboard, the distribution pipes can be run in the floor void and the rest of the service installation can be fixed to surfaces.

A preferable arrangement is to accommodate all pipes and cables in vertical and horizontal service ducts. This produces a neater finish and affords greater ease of maintenance. It is seldom done in private houses, but in buildings of multiple occupancy, the greater complexity of service provision and the need to avoid disturbing one flat to get to the pipes or cables of another has made service ducts a regular feature.

6.7 Service ducts

Ducts for service pipes and cables must run both vertically and horizontally through the building. If the bathrooms and kitchens of the flats are arranged together, a vertical duct can be provided in a central position and can accommodate the drainage pipework as well. Where there is no central service core, the vertical duct may be sited with the stairs and lifts. The horizontal ducts usually follow the route of corridors to allow ease of access, and may be between false ceiling and the structural floor, or formed in the floor itself, below duct covers.

Vertical ducts could constitute a route for fire to spread from floor to floor if the flames can break into and out of the duct enclosure. This must, therefore, be built of a fire-resisting material. The usual construction is of brick (see Fig. 6.3) but a duct casing can be used, provided that it can achieve a period of fire resistance of half an hour. Continuity of the enclosure is essential, so the duct must run from floor to ceiling (or floor to floor if there is a suspended ceiling); and any perforation in the enclosure, to permit the passage of branch pipes, must be as small as is practicable, and must be fire stopped.

Horizontal service runs along the line of a corridor can be

68

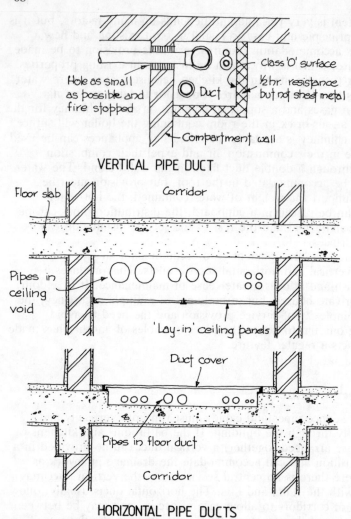

Fig. 6.3 Pipe ducts

suspended from below the structural floor and concealed by a
suspended ceiling, thus using the ceiling void as a duct. In this case
it is necessary to provide a ceiling with access panels or lift-out
ceiling tiles, and the corridor partitions must be carried up to the
structural floor to provide fire protection to adjoining flats (see
Fig. 6.3). To make use of this system the storey height must be
generous, to ensure that the floor-to-suspended ceiling height in the
corridor is adequate. An alternative, not involving extra height in
each storey, is to form a duct in the upper part of the floor. This

can easily be achieved in building blocks where the corridor is centrally placed in the plan and there is a line of structural support (structural walls or beams) along each side of it. In this case the thickness of the corridor floor does not need to be as much as that of the longer spanning floors of the flats and the difference in depth can be used to accommodate a service duct (see Fig. 6.3). The depth available can be increased by dropping the corridor floor slab but this, again, could involve increasing the storey height to leave enough height in the corridor.

6.8 Underfloor heating

A method of space heating which is an integral part of the building fabric is electrical floor warming. In this system, electrical cables, which present a slight resistance to the passage of current, are embedded in the finishing screed (see Fig. 6.4). The slight electrical resistance causes the cable to heat up and this warms the screed, making the whole floor the space heater. Since the surface area of the heater, i.e. the floor, is very large, the temperature difference between the heater and the air can be quite small. Installations are usually designed to maintain a surface temperature of 24°C.

Not only does the screed warm up, but so also does the concrete slab below, thus providing a large heat store. The result of this is a heating system which is very even both in its temperature and in its distribution of heat. Perimeter insulation must be provided, as shown in Fig. 6.4, to prevent excessive heat loss around the edges of the floor. Expanded polystyrene slabs are generally used for this purpose and must be installed right at the beginning of the work, before the concrete floor is laid. A disadvantage of the system is that it is very slow to respond to control because the concrete takes most of the heat when the current is first switched on and continues to radiate heat long after the power has been switched off.

The same system can be applied using low-pressure hot water. In this case the electrical cables shown in Fig. 6.4 are replaced by microbore copper pipes which connect to header pipes across each end, through which heating water is circulated.

One of the implications of floor warming is that the floor covering must be carefully considered for its effect on the heating system and for the effect of the higher temperature of the floor surface on the flooring material. A thick underfelt covered by a deep-pile carpet forms a very effective insulation against heat loss through a floor but, in this case, it would form an equally effective barrier to the space-heating effect of the floor-warming system. If the floor is to be covered by a material which is sensitive to heat or moisture, careful selection of the grade or, in the case of timber,

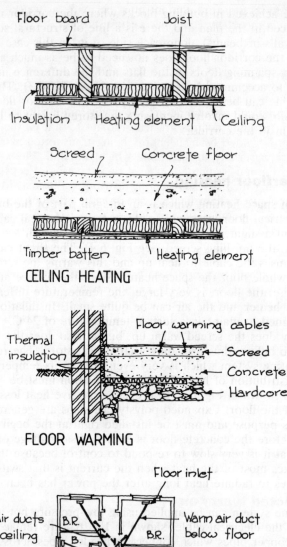

Floor board Joist

Insulation Heating element Ceiling

CEILING HEATING

Screed Concrete floor

Timber batten Heating element

Thermal insulation

Floor warming cables
Screed
Concrete
Hardcore

FLOOR WARMING

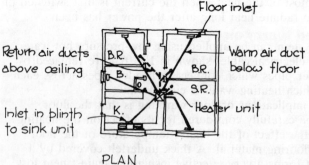

Floor inlet

Return air ducts above ceiling

B.R.

Warm air duct below floor

B.

B.R.

S.R.

Heater unit

Inlet in plinth to sink unit

K.

PLAN

DUCTED WARM AIR

Fig. 6.4 Built-in methods of heating

the species and its moisture content must be made as well as the method of laying and any adhesives to be used.

6.9 Ceiling heating

An alternative to heating the floor is to heat the ceiling. The principles behind ceiling heating are, however, quite different to floor warming. Ceiling heaters are true radiators, that is, their heating effect is gained by heat energy being transmitted as rays which directly warm any surface upon which they fall. Floor warming and normal 'radiators' achieve their heating effect by warming the air which, by convection, distributes the heat to all parts of the room. Radiant heating, as a result, provides comfort conditions with air temperature of only 17°C to 18°C. The lower air temperature gives a fresher feel to the room and there is also a reduction in the amount of heat energy lost through leaks and draughts.

The method of installation is to fix special heating elements to the underside of the joists (if it is a timber floor) or battens (if it is a concrete floor), usually by stapling, connect them to the electrical supply and then to fix a plasterboard ceiling in the normal way. Insulation is laid over the elements to retain the heat in the room for which it is intended. The elements generally consist of glasscloth coated with a silicone elastomer, which conducts the heat, and fitted with copper strip electrodes down each edge to which electrical connections are made (see Fig. 6.4).

The temperature of the surface of the ceiling is relatively high 32°C to 38°C – and to ensure that the heads of the building occupants are not overheated, the ceiling height should, preferably, be more than the minimum of 2.3 m.

6.10 Ducted warm air

Instead of pumping hot water to radiators to heat the air which eventually warms the occupants, it is possible to warm the air directly in a gas, oil or electrical heater and blow it through ducts into the various rooms. Provision must be made at the same time to return the air from the rooms to the heater, otherwise back pressures develop, restricting the flow of the warmed air.

Externally, the heater unit is similar in appearance to a normal boiler (internally it is rather different) and requires the gas, oil or electrical supply running to it and, in the case of the first two fuels, a flue pipe from it. In addition, there are flow and possibly return air ducts to be connected.

Simple systems use a centrally placed up-flow heater with high-level stub ducts into each ground-floor room and a rising duct to the first-floor landing. Grilles in the walls allow the circulating air to return to the heater (see Fig. 6.4). This is inexpensive in its installation but questionable as to its heating standard in all parts of the house. A better system uses ducts, usually of galvanised steel, to distribute the warmed air and, in the most sophisticated systems, to collect it again and return it to the heater.

The ducts have to be accommodated below ground floors, within upper floors and in roof voids, and their size can be quite large: roughly 5000 mm² per kilowatt of heating load, i.e. if the total heat requirement for the house was 10 kW, the maximum duct size at the heater would be 0.5 m² or just over 700 mm square. This, additionally, must be surrounded by insulation to ensure that the heat reaches its intended destination and, consequently, the implications of the accommodation of warm-air ducts are quite considerable. Duct sizes can be reduced from those mentioned above but then the air velocity rises and unacceptable noise develops.

Chapter 7

Energy conservation by the external envelope

7.1 The conservation of energy

In this context, the energy being referred to is heat energy and the object of the conservation is to reduce the quantity of fuel consumed in maintaining comfort conditions within the house. Full comfort conditions depend on more than just the regulation of air temperature. Relative humidity (the amount of moisture vapour in the air), air movement, temperature variation and distribution all contribute to a pleasant internal environment, but it is the maintenance of a standard of warmth which is the principal consumer of fuel and the subject of this chapter.

In medieval days, when the wind howled through the open window holes in the walls, freezing people's backs, and a great fire blazed on a hearth in the middle of the floor, filling the hall with smoke and roasting people's fronts, bodily comfort had to be achieved by clothing. Today, we prefer to dispense with the heavy cloaks, puffed-out garments and voluminous dresses of the past which maintained a comfortable environment close to the body and, instead, to produce the same conditions throughout the whole building so that we can dress in light clothing. Thus, we are particularly concerned with heat losses through that part of the house structure which encloses the space we are endeavouring to keep warm, i.e. the environmental envelope of roof, walls and floor.

Heat energy escapes from all parts of our buildings but this loss is not evenly distributed. Warm air rises and, in any building, it is always hottest at the top of the enclosed space, therefore any roof void is the hottest part of all. Consequently there tends to be the highest rate of loss through the roof but, since the roof represents only a part of the environmental envelope, the proportion of the total heat loss of the building which passes out through this element amounts to about 25 per cent.

The rate of loss through the walls tends not to be so great as through the roof but there is a lot more wall than roof and so the proportion of the total loss is about 35 per cent. The remaining 40 per cent is made up of 15 per cent through the ground floor, 10 per cent through the windows and 15 per cent due to draughts and air leakage.

From the figures, it can be seen that the order of priority when tackling the reduction of heat loss is the walls, the roof, the ground floor or draughts and finally the windows.

That there is a relationship between the heat losses of a building and the heat input is obvious; what is not so obvious is that this relationship is not a simple one-to-one ratio where the amount of heat we must put in exactly equals the amount of heat which gets out. The output of the heat-producing equipment needs only to make up the difference between the heat losses and the incidental heat gains. A building is warmed by the sun falling on the walls and roof and coming in through the windows, by the heat output of the lighting installation, cooking equipment and any other process or equipment which gives off waste heat such as refrigerators, TV sets etc., and by the occupants following their daily activities.

It is unlikely that the incidental heat gains of a building can be altered very much, although careful design can maximise the solar heat gain, but the amount of heat loss is a factor under the control of the designers and builders, and a reduction of this factor increases the value of the heat gains and reduces the amount of heat input needed. There are some buildings where the design and construction have been developed to the point where only on exceptionally cold days do the losses exceed the gains, making it necessary to turn on the heating system.

If it were possible to build in such a way that there were no heat losses through the fabric, the problem of keeping the interior cool would be as great as today's problems of keeping it warm. Indeed, in the highly insulated buildings referred to in the last paragraph, the balance temperature at which the heating system is brought into action has to be set for a higher outside temperature than is really necessary, otherwise the cooling load in the summer would outweigh the heating load in the winter.

There are three ways to prevent heat energy escaping: blanket insulation, reflective insulation and vacuum insulation. The first two are used in buildings but the last, and most effective, is at the moment only feasible for small objects such as the vacuum flask in which liquids can be kept hot or cold.

Reflective insulation can be provided by sheets of bright aluminium foil. It acts in relation to heat rays in the same way that a mirror acts in relation to rays of light, turning back the heat energy which is trying to get out and, incidentally, turning back the heat energy which is trying to get in.

Blanket insulation, the most widely used method, relies on holding a layer of air still. Heat is conducted by the vibration of the molecules in a material, and the more closely packed the molecules, the faster the vibration spreads and the greater the rate of conduction. For this reason, gases conduct heat more slowly than solids. They can, however, carry heat energy by convection, in which the warmed gas moves from the point of heating to a point of cooling. By holding the air still, convection currents are prevented from developing and advantage can be taken of its slow rate of conduction. It is held still by trapping it in pockets which seal in a quantity of air sufficiently small for there not to be any variation in temperature which could cause an air circulation.

A blanket is made up of a mat of fibres between which is held the insulating air. Clothing generally follows this principle: knitted woollen garments contain many air pockets; the string vest, which is full of holes, traps air in those holes between the skin and the next layer of clothing; and the voluminous medieval fashions arose through an unwitting desire to encapsulate even more air. The same applies to building insulation: glass and mineral-fibre quilts are mats of filaments which prevent free air movement within them; expanded polystyrene is even better in that it is composed of completely enclosed bubbles of the plastic in which any air movement is totally impossible.

It should be noted that the thermal conductivity of the material used to form the pockets containing the air has relatively little bearing on the insulating quilt or board. Glass and polystyrene have high thermal conductivities and the work of drawing the glass into filaments or expanding the polystyrene into nodules does not change the ability of the material to conduct heat. However, when the filaments are interwoven into a quilt or the nodules made up into a board, the final result is a building material of high resistance to thermal transmission.

In any consideration of the conservation of heat energy, the shape as well as the construction of the building must be taken into account. Heat is lost from the enclosed volume through the enclosing fabric: the smaller the area of fabric compared to the volume enclosed, the less the heat loss irrespective of the thermal conductivity of the envelope. The two building shapes shown in Fig. 7.1 both contain 162 m³ of enclosed space, but building A has only 198 m² of external surface, whereas building B has 234 m² of wall and roof and, consequently, a higher heat loss. Building A would benefit further because it has more of the enclosed volume

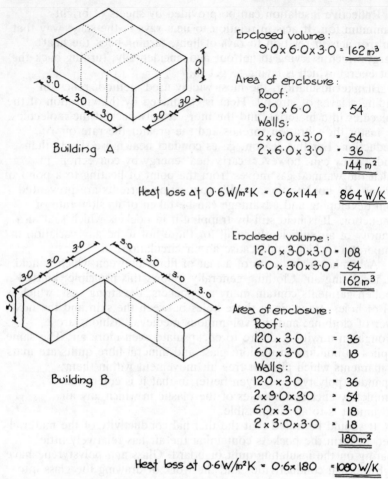

Enclosed volume:
$9.0 \times 6.0 \times 3.0 = 162\,m^3$

Area of enclosure:
Roof:
$9.0 \times 6.0 \qquad = 54$
Walls:
$2 \times 9.0 \times 3.0 = 54$
$2 \times 6.0 \times 3.0 = 36$
$\underline{144\,m^2}$

Heat loss at $0.6\,W/m^2K = 0.6 \times 144 = \underline{864\,W/K}$

Enclosed volume:
$12.0 \times 3.0 \times 3.0 = 108$
$6.0 \times 3.0 \times 3.0 = 54$
$\underline{162\,m^3}$

Area of enclosure:
Roof:
$12.0 \times 3.0 \qquad = 36$
$6.0 \times 3.0 \qquad = 18$
Walls:
$12.0 \times 3.0 \qquad = 36$
$2 \times 9.0 \times 3.0 = 54$
$6.0 \times 3.0 \qquad = 18$
$2 \times 3.0 \times 3.0 = 18$
$\underline{180\,m^2}$

Heat loss at $0.6\,W/m^2K = 0.6 \times 180 = \underline{1080\,W/K}$

Fig. 7.1 Effect of building shape on heat loss

remote from the envelope, losing no heat at all. The ultimate development of this principle is a spherical building, since this shape has the minimum surface area in relation to the volume enclosed but, of course, it is not a practical proposition.

Precisely what ambient temperature represents the ideal standard of warmth varies from person to person and from time to time. Therefore, any table of recommended temperatures gives the average acceptable figure for the majority of the population. If it is known that the occupants will all be of a particular social group, i.e. elderly people, these standards will be varied to suit the average requirements of the group. In the case of elderly people, the general temperature is higher. The factors which determine the ideal personal temperature are sex, age, size, activity and clothing and each can create wide variations in the desired level of heat. Fortunately, the human body has the ability to make adjustments to deviations from the ideal temperature, by shivering or sweating, and we also adjust the amount of clothing we wear to suit the activity to be carried on.

The tendency is for heating standards to increase steadily, in keeping with standards of domestic comfort and convenience generally. In parts of America, standards are higher than in the UK where the current recommended domestic temperatures are:

Living rooms   20–21 °C
Bedrooms   13–16 °C
Kitchens   16 °C

Although all ages of occupant may prefer a departure from these standards (some people like hot bedrooms), it is only in housing for the elderly where a different standard is recommended and, in this case, the building should maintain a temperature of 21 °C throughout the complete enclosed space.

7.2 Measurement of heat loss

Apart from the 15 per cent lost through draughts and air leakage from the building, heat is lost by direct transmission through the materials of the enclosing envelope. To be able to know, in advance of the need arising, the demands which will be placed on the heating system, it is necessary to calculate the total potential heat loss of the building using known values of thermal conductivity.

It is possible, by laboratory tests, to find out how much heat a particular material can conduct. The amount depends not only on the characteristics of the material, but also on its thickness and on the difference in temperature between the two faces. To make it possible to compare materials and to bring calculations to a common basis, the last two factors are standardised at 1 m thickness and 1 K (temperature differences should be expressed in Kelvin not degrees Celsius). Thus common brickwork has a thermal conductivity (k value) of 0.84 W/m²/m/K. This means that if a solid brick wall is 1 m thick and that one side was 1°C warmer than the other side, there would be a flow of heat energy through each square metre of wall face of 0.84 watts. In practice, the units are cancelled out and k values are quoted in W/mK.

For convenience in calculating potential heat loss, it is more usual to use the resistivity (r value) of the material, which is found by dividing the k value into 1.0. For the common brickwork quoted above, this r value is 1.19 mK/W (the units are reversed to indicate the reciprocal). By multiplying the resistivity (r) of a material by its thickness, the thermal resistance (R) is found. The resistance of 225 mm of common brickwork is, therefore, $1.19 \times 0.225 = 0.26$ m^{2}K/W.

This, however, is the resistance of the material, not of the wall, because the surfaces offer a resistance independent of the material. Surface resistance to the net transfer of heat depends on its emissivity (its relative ability to radiate heat), its absorptivity (its ability to absorb heat energy) and its reflectivity (its ability to reflect heat). The value is also affected by the temperature and speed of any wind across its surface. Since no thickness is involved, these surface resistances are expressed in m^{2}K/W and for wall faces they are 0.123 m^{2}K/W internally and 0.055 m^{2}K/W externally where exposure is normal. See *Building Research Station Digest 108* for the full range of values.

The value the heating engineer requires to know is the overall rate of transmission for any given structure. This overall rate or air-to-air heat transmittance coefficient is found by adding up all the resistances and dividing the total into 1.0. Thus for a 225 mm brick wall it would be:

Internal surface   Brickwork   External surface

$$0.123 \quad + \quad 0.26 \quad + \quad 0.055 \quad = 0.438 \text{ m}^2\text{K/W}$$

$$\frac{1}{0.438} = 2.28 \text{ W/m}^2\text{K}$$

This result means that, in the wall under consideration, 2.28 watts of heat energy passes out of every square metre of wall for each degree of temperature difference between the faces. This is known as the U value.

Had there been a cavity in the wall, this would also have presented a resistance to the passage of heat, and in normal cavity wall construction the value of the resistance is 0.18 m^{2}K/W. If, therefore, the brick wall already calculated was divided into two half-brick skins with a 50 mm cavity, its total resistance would rise to $0.438 + 0.18 = 0.618$ m^{2}K/W and its U value would fall to 1.62 W/m^{2}K.

Applying plaster to the internal face adds another resistance; in this case the k value is 0.5 W/mK so its r value is 2.00 mK/W. If its thickness is 19 mm and its resistance is 0.038 m^{2}K/W, the total resistance for the wall increases still further to 0.656 m^{2}K/W and the U value drops to 1.52 W/m^{2}K, indicating that a plastered cavity brick wall is a better thermal insulator than an unplastered solid brick wall.

The same calculations can be applied to the roof, the doors, the windows and any other construction included in the enclosing envelope to find the U value for each. By multiplying these U values by the total area of the element to which they relate and adding all the results together, a total heat loss figure is found which is used in deciding the output capacity of the heating system.

To save the heating engineer having to carry out repetitive calculations of U values for each building on which he works, tables have been published by the Chartered Institute of Building Services and in *B.R.E. Digest 108*, giving the thermal properties of building materials and constructions including the U values for a range of common forms of wall and roof construction.

7.3 Building Regulation standards

The standard of thermal insulation which must be achieved by the external envelope of a house is now governed by the Building Regulations but it has been a statutory requirement in England and Wales since the issue of the Model Byelaws in 1953 which set a standard roughly equivalent to a brick cavity wall.

The first Building Regulations set maximum U values for walls as 0.30 BTU/ft²/hr/°F (1.70 W/m²K) and for roofs as 0.25 BTU/ft²/hr/°F (1.42 W/m²K). The walls of rooms in roofs had to possess the same value as walls generally, but any sloping soffits were subject to the standard for roofs. Where a floor was exposed to external air on its underside, e.g. in a bedroom over an open car port, it also had to possess a thermal insulation of the same standard as for roofs, i.e. a U value of 1.42 W/m²K.

In 1972 a new set of Building Regulations was published which incorporated all the seven amendments to the 1965 Regulations. Between 1972 and 1976, three amendments to the 1972 Regulations were issued, the second of which completely revised the original requirements by substituting higher standards of insulation generally plus an overall U value for walls when losses through windows, doors and other openings were included. These new standards set a maximum U value for any part of a wall as 1.0 W/m²K, but added that the averages for all the perimeter walling (including a party wall) was not to exceed 1.80 W/m²/K. The U value for roofs was also reduced to 0.6 W/m²K.

The three amendments were consolidated into another set of Regulations, the Building Regulations 1976. These were subsequently the subject of amendment and in 1981 the second amendment raised further the standard of insulation by setting lower U values of 0.6 W/m²K for walls and 0.35 W/m²K for roofs. These standards still apply. As well as lowering the U value, the 1981 amendment also departed from the idea of an overall average U value and substituted limits of permitted areas of windows. These were expressed as percentages of the total perimeter walling and are:

Single-glazed windows   12%
Double-glazed windows   24%
Triple-glazed windows   36%

If the actual standard of insulation of the wall exceeds the minimum laid down, these areas may be increased proportionally. Recognising that with good standards of thermal insulation comes a high risk of condensation, the second amendment 1981 required that pitched and flat roofs must be ventilated. Figure 7.2 sets out

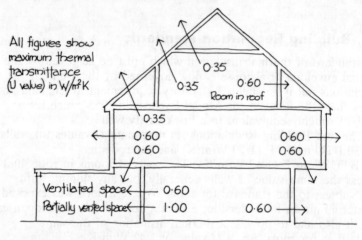

All figures show maximum thermal transmittance (U value) in W/m²K

0·35
0·35   0·60
Room in roof
0·35
0·60   0·60
0·60   0·60
Ventilated space   0·60
Partially vented space   1·00   0·60

THERMAL INSULATION

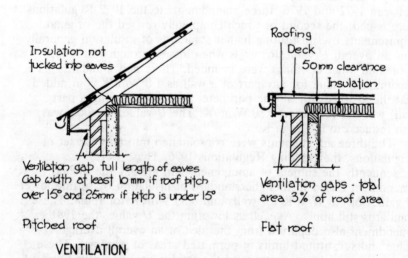

Insulation not tucked into eaves

Ventilation gap full length of eaves. Gap width at least 10 mm if roof pitch over 15° and 25mm if pitch is under 15°

Pitched roof

Roofing Deck
50mm clearance
Insulation

Ventilation gaps - total area 3% of roof area

Flat roof

VENTILATION

Fig. 7.2 Thermal insulation and roof ventilation regulations

diagrammatically the complete thermal insulation standards and ventilation requirements currently applicable.

7.4   The insulation of roofs

A pitched roof of normal construction comprising a timber framework covered with felt and tiles and lined with plasterboard has a U value of about 2.0 W/m^{2}K. The Building Regulation standard is 0.35 W/m^{2}K and, therefore, some thermal insulation must be introduced.

There are two locations for thermal insulating material, over the rafters or on the ceiling. Placing the insulation over the rafters was a favoured practice by some builders when standards were lower but now, to achieve the thermal resistence needed, a thickness of insulation of 90 to 100 mm is needed. This makes it difficult to obtain a solid fixing for the tile battens. The advantage of over-rafter insulation is that if keeps the roof space warm and reduces the risk of freezing pipes or tanks. It could also be shown to save costs since it could be applied with the roofing felt in one operation. The usual practice, at the present time, is to lay a 100 mm thickness of glass fibre or mineral wool between the ceiling joists directly on to the plasterboard. For many roofs, 80–85 mm thickness of glass fibre would achieve the minimum thermal resistance, but the material is most commonly available in the thicker size. Figure 7.3 gives minimum thickness for a range of insulating materials.

Two hazards arise with ceiling insulation: condensation in the cold roof and freezing of water pipes and tanks. The latter is prevented by insulating the pipes and tanks and omitting the ceiling insulation below the tanks while the problem of condensation is dealt with by ventilating the roof space (see Fig. 7.2).

Uninsulated flat roofs allow more heat to escape than uninsulated pitched roofs and, therefore, the incorporation of a thermal insulant is even more necessary. The required U value can be achieved by laying glass fibre on the ceiling (Fig. 7.3A), on the roof deck (Fig. 7.3B) or on top of the roofing (Fig. 7.3C).

The insulated ceiling method produces a 'cold' roof in that there is a void between the insulation and the roof deck which is unheated. The Building Regulations require this void to be ventilated as shown in Fig. 7.2, and BS 5250 : 1975 recommends a minimum gap of 50 mm between the insulation and the roof deck.

The other two methods produce a 'warm' roof, i.e. the roof void is heated by the air rising from the room below. In this case, the integral vapour barrier is essential to prevent condensation within the deck or insulation. Since there is no void above the insulation, the Building Regulation ventilation requirement does not apply, nor is it sensible to pass external air through a 'warm' roof

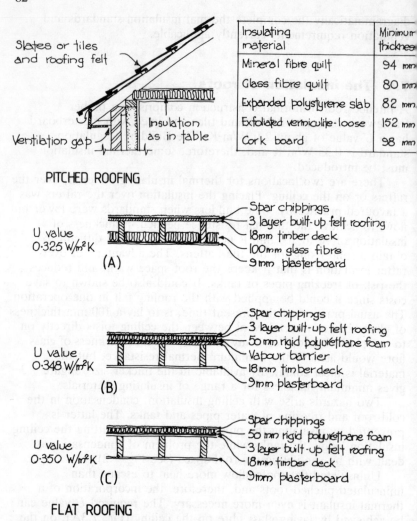

Insulating material	Minimum thickness
Mineral fibre quilt	94 mm
Glass fibre quilt	80 mm
Expanded polystyrene slab	82 mm
Exfoliated vermiculite-loose	152 mm
Cork board	98 mm

Slates or tiles and roofing felt

Ventilation gap

Insulation as in table

PITCHED ROOFING

U value 0·325 W/m²K

(A)

- Spar chippings
- 3 layer built-up felt roofing
- 18mm timber deck
- 100mm glass fibre
- 9mm plasterboard

U value 0·348 W/m²K

(B)

- Spar chippings
- 3 layer built-up felt roofing
- 50 mm rigid polyurethane foam
- Vapour barrier
- 18mm timberdeck
- 9mm plasterboard

U value 0·350 W/m²K

(C)

- Spar chippings
- 50 mm rigid polyurethane foam
- 3 layer built-up felt roofing
- 18mm timber deck
- 9mm plasterboard

FLAT ROOFING

Fig. 7.3 The insulation of roofs

because this would increase the rate of heat loss. The warmed air passing up into the roof voids from the room below is usually moist and to protect the timbers from any adverse effects. BS 5250 recommends that they should all be treated with a pressure-impregnated preservative.

7.5 The insulation of walls

Walls present the same problems as roofs, in that traditional

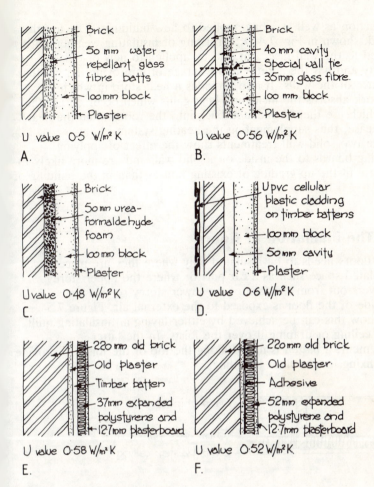

Fig. 7.4 Wall insulation

construction does not achieve the stipulated national standards of thermal insulation. Therefore, additional provision must be made. Figure 7.4 shows six typical insulation arrangements, the first four of which, A, B, C, and D, apply to cavity construction, while E and F are based on a solid brick wall.

Illustrations A and B of Fig. 7.4 show two methods using glass fibre or mineral wool built into a cavity wall as it is constructed. Where the cavity is completely filled, the insulation must be moisture resistant, to withstand the effects of rain penetrating the outer leaf. Alternatively, the cavity can be filled with an insulant after it has been built. A variety of suitable materials are available, and Fig. 7.4 C shows the value for urea-formaldehyde foam. This method can be used to upgrade existing property of cavity

84

construction as well as being applied to new buildings. The fourth
method, shown in Fig. 7.4 D, consisting of insulating weatherboard
attached to the wall face, can also be applied to new or old
property and has the additional advantage of improving the weather
resistance of the walls. It also provides a heat store in which the
brickwork absorbs the incidental heat gains which occur during the
day, which are then re-radiated back into the room when the heat
gains cease, thus supplementing the heating system.

The two solid-wall treatments show the effect of applying
insulating boards to the inside of a solid wall and are more likely to
be of use in the up-grading of existing houses than in the building of
new.

7.6 The insulation of floors

Upper floors needs to be insulated only where they are placed over
a ventilated space such as a car port or where the upper storey
cantilevers out from the face of the lower storey so that the
underside of the floor is exposed to the external air. Figure 7.5
shows how this can be achieved by either laying an insulating quilt
on the ceiling or draping it over the joists. In the latter
arrangement, a batten is nailed along the top of the joists to provide
a firm fixing for the floor boarding.

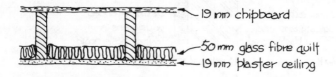

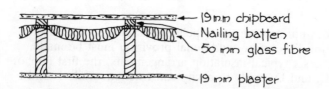

U value 0·55 W/m² K

Fig. 7.5 Thermal insulation of floors

Chapter 8

Energy conservation in window design

8.1 Heat loss through windows

In the last chapter, consideration was given to the forms of construction for walls and roof which would achieve acceptable levels of thermal insulation. The total heat loss is not only the product of the thermal transmission through walls and roof, but is also related to the perforations in those elements in the forms of doors, windows, skylights etc. On average, some 25 per cent of heat loss is attributed to openings in the envelope, made up of 10 per cent direct loss and 15 per cent due to draughts and air leakage. Naturally, these figures vary considerably between one building and another, and are related to the total area of window and door, its orientation and its detail design.

Obviously, the more window area there is, the more heat loss occurs, since the thermal insulation standard of a window cannot equal that of the wall it replaces. To control this, as explained in Chapter 7, the Second Amendment (1981) to the Building Regulations 1976 introduced the concept of defining maximum percentage area of glazing in relation to the area of enclosing structure. For houses, this was set at 12 per cent for single-glazed windows, 24 per cent for double-glazed and 36 per cent for triple-glazed. While this simple rule controls the effect of openings on the overall heat loss, it does not recognise that windows on the north side of the building lose more heat than those on the south because of the greater temperature difference between inside and outside.

Furthermore, south-facing windows can contribute significantly to the heating of the building by allowing the sun to heat the floor and objects within the room to produce an incidental solar heat gain. It follows, therefore, that any building, no matter what its construction, will require less heating if it has only a small amount of north-facing windows, or none at all, and generous south-facing openings. These latter openings should also be equipped with heavy curtains to counteract the potentially large heat loss when the external temperature drops at night.

8.2 Paths of heat loss through a window

As already stated, window heat loss is due to direct transmission and also to air leakage. Direct transmission occurs through the glass and also through the frame. Transmission through the glass can be reduced by double glazing (see sections 8.3 and 8.4), and through the frame by a thermal break (see section 8.5). Thermal breaks are more important in metal frames, with their high thermal conductivity, than they are in timber frames.

Air leakages, or draughts, occur wherever there is a crack or opening in the frame. There are two places where these cracks can be found: around the edges of opening lights and between the main frame and the adjoining structure. The closing of gaps around the opening lights is achieved by weather stripping and around the frame by careful installation to ensure that the joint is well flushed up with mortar or caulked with mastic (see section 8.6).

8.3 Double glazing

In Chapter 7 it is explained that, in most building techniques, still air is the insulating medium used. Double glazing relies on the same principle. The transmission of heat through the glazing is reduced by using two sheets of glass which trap a layer of air between them. In the case of blanket insulation, the air is held still by a mat of fibres, but this cannot be used in glazing. Instead, movement of the air is restricted by careful selection of the distance between the sheets of glass. Naturally, the more air that is held still the better the insulation value, but if the thickness of the layer of air is too great, an air flow can be set up, by convection.

In this event the air rises up the face of the inner pane of glass, absorbing heat as it goes, crosses the air space at the top and drops down the face of the outer pane, losing its heat to the cooler external glass as it descends. This almost completely cancels out the potential benefit of the double glazing.

Investigation has shown that the optimum width of air space is

about 20 mm but, since the thickness of a double-glazing unit
containing this optimum space would present practical problems with
window frame thickness, it is reduced, with only a small loss of
efficiency, to between 5 mm and 12 mm. At these spacings the
reduction in thermal transmission, in comparison to single glazing
(taking the U value for single glazing as 5.6 W/m²K), is as shown
below.

Air space (mm)	U value W/m²K	Reduction in transmission (per cent)
20	2.9	48.21
12	3.0	46.42
5	3.5	37.50

In addition to keeping the trapped air still, it is advantageous to
dry it to prevent condensation forming on the inside face of the
outer pane of glass.

Sealed units are one of the most common methods of providing
double glazing and consist of two sheets of clear glass, or one clear
and one obscured, which are hermetically sealed to a metal spacer
along each edge. The air between the glass is withdrawn and
specially dried air is introduced. The unit is finished with a
protective metal tape enclosing the edges. With two sheets of glass
and an air space, a sealed unit takes up more of the glazing rebate
in the frame than a single pane of glass. To accommodate this, the
rebate must be made deeper and the frame thicker.

Alternatively, the rebate and frame can be left as for single
glazing and a stepped double-glazing unit used. Stepped units have
one pane of glass larger than the other which leaves a protruding
single thickness of glass to be glazed in the normal way (see Fig.
8.1). When glazing, the units must be placed on setting blocks to
maintain the specified glazing clearance, and the unit secured with
glazing beads and non-setting glazing compound. Two other methods
of double glazing are twin frames and dual glazing (see Fig. 8.1). In
both cases the air space is not hermetically sealed and, therefore, it
is necessary to install tubes to allow the air to 'breathe' to the
outside air to prevent condensation which slightly reduces the
insulation value. It is also necessary to clean the glass facing into
the air space periodically and provision must be made in the design
for suitable access.

8.4 Secondary glazing

To reduce heat loss, many existing house windows are now being
fitted with a system of secondary glazing. In a lot of cases this is
fitted by the householder. Details differ between manufacturers, but

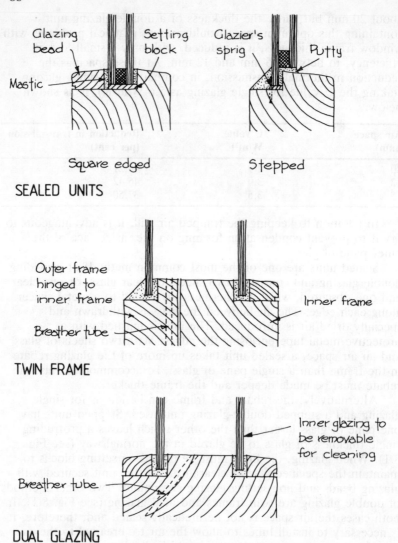

Fig. 8.1 Double-glazing methods

most comprise an aluminium or plastic frame fitted on to a sheet of glass and attached to the inside of the existing window with clips or turn buttons, or an aluminium-framed horizontal sliding sash fitted to a sub-frame, fixed to the inner window reveal. The design allows for the secondary glazing to be released or opened either to allow the main window to be opened or for cleaning purposes (see Fig. 8.2).

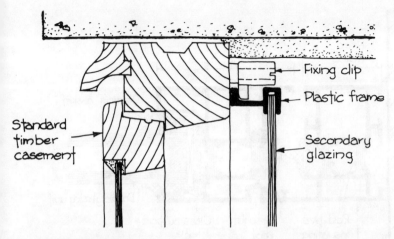

Fig. 8.2 Secondary glazing

The advantages of secondary glazing are that it is a simple and inexpensive way to upgrade the standard of thermal resistance of an existing window and, since the inner sash usually completely covers the opening lights, it also reduces draughts. On the other hand, the width of the air space is, inevitably, quite large and possibly sufficient to allow a convection current to start which will eliminate much of the potential value of the secondary glazing. It is also difficult to control condensation forming within the air space.

8.5 Thermal breaks

Metal window frames, with their high thermal conductivity, can create a substantial heat loss and the cold surface of the metal can cause condensation to form in sufficient quantities to disfigure the decorations of the window reveal as well as leaving pools of water on the window cill.

The only satisfactory way to reduce the high energy transmission through the frame is to introduce a thermal break. This is done by forming the frame in two halves between which is sandwiched a pad of insulating material so arranged that there is no metallic contact at all between the inner and outer halves of the frame (see Fig. 8.3).

8.6 Draught exclusion

As already mentioned, there are two routes for air leakage at a window, at the opening light and between the frame and the enclosing structure.

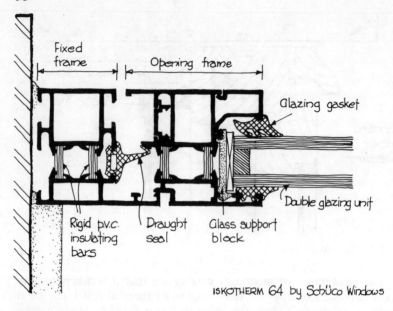

Fixed frame

Opening frame

Glazing gasket

Double glazing unit

Rigid p.v.c. insulating bars

Draught seal

Glass support block

ISKOTHERM 64 by Schüco Windows

PLAN

Fig. 8.3 Aluminium window with thermal break

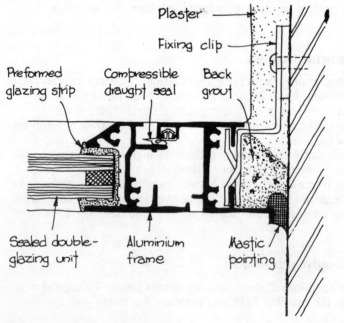

Plaster

Fixing clip

Preformed glazing strip

Compressible draught seal

Back grout

Sealed double-glazing unit

Aluminium frame

Mastic pointing

Fig. 8.4 Draught exclusion

Draughts from between the opening light and the frame can be prevented by compressible or wiping seals. The type of seal required depends on the nature of the relative movement of the two parts of the frame. Compressive seals, usually of an elastomeric plastic, are either U-shaped or tubular and are flattened or distorted by the closing action of the frame. This causes them to fit tightly to the closing surfaces, sealing the opening against air movement. Wiping seals, with either an elastomeric tongue mounted in an aluminium channel or a nylon brush welt, are used in conjunction with surfaces that slide relative to each other.

Figure 8.4 shows the plan through the jamb of an aluminium window incorporating a compression seal on to which the opening light closes. The drawing also shows the backing grout and mastic pointing employed to seal the frame to the adjoining wall reveals.

Chapter 9

The control of domestic building construction

9.1 The need for control

In any civilised society it is necessary to impose codes of conduct to preserve the community and to protect its members, generally, against harmful acts by individuals. As the community's life becomes more complex, the ways in which one individual can harm another grow more subtle and can be committed quite unwittingly. One of the greatest impacts an individual can make on a community is to erect a building. In doing so to his maximum advantage, he can rob other members of the chance to use their property as they wish. Any developer wishes to put his time and money to the best use, but if he is allowed to work unchecked he could, quite innocently, create discomfort, deprivation or danger to others. The way this risk is minimised (it is very difficult to avoid a new building having some adverse effects) is for the community to empower its governing body to impose rules of building. In this country that body is central government and the rules are the Building Regulations.

9.2 The development of control

Throughout history there have been codes or rules for building, but none as complex or comprehensive as the UK Building Regulations. The Romans, for instance, adopted the simple measure that if a building collapsed and killed someone, they executed the builder! More positive methods are now used.

Control in England really started in the middle of the seventeenth century and arose, as is often the case, after a series of catastrophes, in this case fires. There were many serious fires at this time which consumed large areas of towns and cities – Nantwich 1583, Bury-St-Edmunds 1608, Dorchester 1613, Great Fire of London 1666, Northampton 1675 and Warwick 1694. It was the devastation of London which prompted the first rules of building, intended, not surprisingly, to prevent the spread of fire. From these rules, and the 'discreet men knowledgeable in building' who were appointed to enforce them, can be traced the present London Building Acts and the London District Surveyors. These rules were followed by a number of pieces of legislation which had the objective of controlling house building, notably the Public Health Act of 1875. This gave local authorities the power to make local bylaws to impose standards in relation to safety, fire prevention, health and sanitation in houses. The system produced widely varying requirements and standards, despite the guidelines of a set of model bylaws produced by the government. There were about 1400 local authorities and the same number of different sets of bylaws, which made life very difficult for designers and builders.

In 1961, a new Public Health Act empowered the making of national Building Regulations, the first of which came into operation in February 1966. There then followed seven amendments which were consolidated into a completely revised set of Regulations dated 1972. With further developments in building techniques and building control policy, more amendments were needed in 1978, 1981 and 1983, following which it was decided to change the format of the way the Regulations were produced to the three volumes of Regulations, Manual to the Regulations and Approved Documents in which are incorporated all three amendments.

Unlike the earlier building codes which were laid down, the control of building now uses performance requirements as the criteria so, instead of stating how a wall should be built, the newer Regulations require that, when built, a wall will achieve a certain standard of performance; thus leaving the designer free to devise his own construction. For guidance, ways of conforming to the standards are out in the Approved Documents.

Originally, building control was imposed only on housing, partly because, at the time, there were predominantly more houses than any other building and economic pressures were depressing standards alarmingly, and partly because other buildings were controlled by specific legislation such as the Factories Act. In recent years, the breadth of control by the Building Regulations has increased to cover non-domestic buildings. It is the declared policy of the government to embrace all legislation concerning building into one single Act.

9.3 The control of aspects of building design

In imposing standards on built form in this country, it is insufficient
to control just the methods of construction. This can result in a
structure which is soundly built but is dangerous or inadequate in
other respects, such as the steepness or protection of stairways, the
amount of natural light and ventilation enjoyed by the occupants,
the height of the rooms, or the means of escape in the event of a
fire.

9.3.1 Staircases

Of all the elements in a building, the staircase is the one which has
most closely to match the size and capacity of the human body. It is
also potentially hazardous because it is concerned with vertical
travel. There are, then, two main areas to impose controls to ensure
a satisfactory stairway, the ergonomic aspects of the efficiency with
which we can ascend the stairs and the anthropometric aspects of
the sizes of the parts which will allow us to traverse the stairs in
comfort and safety.

The dimensions of each step must fall within very close limits.
The height of the riser must be such that it is within the capacity of
90 per cent of the population to raise themselves from one tread to
the next and yet it must be great enough to make the rate of
progress commensurate with the effort expended. The depth of the
tread, i.e. its dimension in the direction of the flight of stairs, must
be large enough to accommodate the average foot, but not so large
that an imbalance is created between the energy we must consume
in lifting ourselves up from one tread to the next and the amount
used in travelling forward. This is controlled by rules governing the
maximum and minimum dimensions of the rise and the going (the
horizontal distance between the nosing, or front edge, of one tread
and the nosing of the one above), the ratio between them and the
angle of a line touching the nosings (the pitch angle).

Not all staircases are straight: some follow a curve in part or for
the full length and, therefore, contain treads which are tapered. To
maintain the safety of people using the stairs it is necessary to set
additional standards for this situation, controlling the angle of taper
and the minimum depth of tread.

In all cases, consistency is most important, so every rise in every
flight of one staircase must be the same, every tread must have the
same depth, and tapered steps must maintain a constant going and
rate of taper.

The safety aspects of stairs are concerned with falling off or
through the staircase or down the stair well. Falling off the staircase
is prevented by a balustrade which is topped by a handrail placed at
840 to 900 mm above the pitch line as an aid to the ascent of the
stairs. Falling through is prevented by the elimination of all

openings in the balustrades or staircase through which a 100 mm sphere could be passed. This usually makes it necessary to introduce some feature to reduce the space between treads in an open riser stairs. Protection against falling down the stair well is provided by a guard rail fixed at 1.1 m above the upper floor and a balustrade between the floor and the rail.

Control is also imposed on other aspects of staircase design, such as the width of the stairs, the number of steps in one flight (for the benefit of the elderly and those who suffer from vertigo) and the clear space above the stairs to provide adequate headroom etc., and also on the design of ramps and stepped ramps.

9.3.2 Natural light and ventilation

These Regulations are mainly concerned with windows and, to a lesser extent, doors. The Regulations impose very little control over natural light, merely the provision of an open space outside the windows of a habitable room, but they do require a specific amount of ventilation provision. Indeed, there is no obligation to provide windows at all and, therefore, one could leave the walls completely blank and rely entirely on artificial lighting and mechanical ventilation. For most people this would be quite satisfactory from a health and sanitation point of view, but for the unfortunate minority who suffer from claustrophobia, it would create quite unacceptable conditions.

9.3.3 Height of rooms

That the height of rooms should be adequate is an obvious requirement of good building design; that it should be necessary to impose control in this respect indicates the appalling state of affairs early building legislation set out to correct. What, exactly, is 'adequate' is defined, currently 2.3 m, and how it is to be measured in varying situations is also determined by the Regulations.

9.3.4 Means of escape

Satisfactory means of escape in the case of fire is the subject of relatively recent legislation and, while it is a complex and extensive subject, it is dealt with simply in the Regulations by requiring reasonable provisions to allow occupants to reach a place of safety. The way by which this is achieved is the subject of other publications, such as *B.R.E. Codes of Practice*.

9.4 The control of structural stability

A fundamental requirement for any building to fulfil its function is that it shall remain stable under all conditions of use throughout its life. This stability stems from two sources, the foundations and the structural elements of the building.

The purpose of foundations, and therefore, the subject of building the control, is to receive the loads of the building and to transmit them to the ground in such a manner that no movement occurs which could cause damage to the superstructure. A further requirement is that the foundations must resist any attack by subsoil water or salts.

There are two aspects to foundation performance: firstly, the transference of the loads in such a manner that the forces imposed on the subsoil remain within its bearing capacity and, secondly, that this transference takes place at a depth where the subsoil is unaffected by changes in climatic conditions. Both of these aspects are the subject of Regulations.

The forces on the superstructure and the methods of construction are more complex than those connected with foundations, but the functional requirements are quite simple. The loads to be dealt with are the dead load of the structure itself, the live loads imposed by the occupants and the wind loads. How these are to be calculated and the values to be taken are defined in the Regulations. With regard to the methods of construction, the functional requirement is simply that the structure will safely sustain the loads and transmit them to the foundations without suffering deformation. How this is to be achieved is the subject of Codes of Practice to which the Regulations refer.

Following the Ronan Point disaster in which an explosion initiated the progressive collapse of the corner of a tall block of flats, the Regulations also require that any building of five or more storeys must be designed so that if a structural member is removed, consequential structural failure is restricted to just the storey immediately above or immediately below.

9.5 The control of fire resistance

Statutes concerned with the effects of fire in a building are the oldest of our building control measures. In some cases they are still in evidence, as the party walls of London terrace housing, protruding above the roof slopes, bear witness. This was a rule imposed in 1667, following the Great Fire of London. An even earlier ruling was one by William the Conqueror who decreed that all domestic fires must be covered at a specified hour at night. The fire cover or 'couvre-feu' gave us our present word of 'curfew'.

The objectives of present building legislation are to prevent a fire in one building spreading to others and to control the effects of the fire within the building itself.

Fire spreads by either passing directly from the building which is alight to any which connect to it, or igniting adjoining buildings by heat radiated through windows, doors or any combustible parts of the enclosing envelope which have collapsed.

In all but houses in single occupancy, the Regulations require any fire to be confined to the part of the building where it started, by the introduction of fire-resisting separating walls and compartment walls and floors. These must withstand the effects of a fire for a defined period of time and any perforations must be sealed with suitable doors, fire stops or other specialised constructions. To enable the occupants to escape from the fire-confining compartments, stairways must be enclosed in a protective shaft.

Ignition by radiation heat is controlled by defining the ratio between the size of the openings, referred to as unprotected areas, and the distance between adjacent buildings. Since this must be applicable to any building regardless of the proximity of adjoining buildings, it must be assumed that the neighbouring structure is built right up to the boundary.

The ratio is, therefore, related to this boundary line. However, the side of a building facing a street or river must, of necessity, be kept remote from any wall facing it by the intervening feature. In this case, the ratio is related to the distance to the centre of the road or river. The principle of control is to assess the level of hazard created by the area of the openings and to adjust the minimum distance between the building and the boundary according to the degree of danger.

9.6 The control of weather resistance and damp

To reach a satisfactory standard of functional performance, a building must withstand, for the duration of its life, the effects of moisture in every direction – snow and rain on the roof and walls, and rising moisture below the ground floor. The Building Regulations deal with this by requiring that such moisture is prevented from entering the building. More precisely: ground floors must prevent the passage of moisture to the upper surface, walls must not transmit water from the ground and must resist the penetration of the moisture due to rain or snow and roofs must be weatherproof.

These requirements are met by the use of damp-proof membranes in floors, damp-proof courses and cavities in walls and sound coverings to roofs laid at the correct pitch.

9.7 The control of thermal insulation

As mentioned in Chapter 7, heat loss from buildings is now a matter of national concern and must be restricted in the interests of the conservation of energy resources. This is a departure from the former principles behind building control – the health and safety of the building's occupants – and imposes a control which has more to do with the interests of the nation than the benefits to the individual. Nonetheless, the occupant does enjoy a greater degree of comfort and smaller bills for his heating.

Control is exercised by defining a maximum value for the coefficient of thermal transmission (U value) for walls and roofs. It is then up to the building designer to either devise and justify his own method of construction or adopt an approved standard method.

Windows increase the overall heat loss by substituting an area of greater transmission for the better insulating wall. To limit this effect, the area of window is restricted to 12, 24 or 36 per cent of the wall area (see Ch. 8).

Two other aspects of former energy wastage now controlled by the Building Regulations are those of inadequate control of heating systems, and heat losses from pipes, ducts and storage vessels. In neither aspect does this control apply to houses, flats or maisonettes. The Regulations now require thermostats to be fixed inside and outside the building to regulate the output of the system and adequate thermal insulation on pipes and storage vessels.

9.8 The control of sound insulation

This, like the Thermal Insulation Regulations, is a relatively recent subject for control and deals with an aspect of building which, although conducive to greater comfort for the occupants, is not of such a high importance as a weatherproof roof or stable foundations and thus did not concern the early legislators. Control is restricted to the construction of walls and floors separating habitable rooms in one dwelling from another dwelling or other buildings.

Noise nuisance can be due to either airborne sound or impact sound. The former travels in all directions and, therefore, dictates the construction of walls and floors above or below the dwelling, while the latter only influences the design of the floor of a dwelling above another, but is the more difficult to contain. In both cases the Regulations require an adequate resistance to the transmission of sound.

Constructions considered adequate in their resistance to airborne or impact-sound transmission are required to possess a specified mass which will absorb a large proportion of the sound energy. Impact-sound insulating floors have, in addition, a soft absorbent layer incorporated, either as the floor covering or as a sound-deadening quilt within the construction.

Part B

Commercial and industrial construction

Chapter 10

Forms of construction

10.1 Factors influencing the choice of superstructure

Many considerations are involved in the selection of the type and method of application of a structural system for a particular office or factory. Of these, the main factors are: whether it is to be a single- or multiple-storeyed building; whether the interior is to be a clear open space or subdivided into rooms; the loads to be carried; the availability of resources; constraints of site and access; the desired aesthetic appearance both externally and internally; and, finally, cost. All these exercise an influence both individually and in their relationship to each other.

The question of single or multiple storeys determines whether suspended floors must be incorporated within the structure. These impose loads on the vertical elements of the building and also, usually, make the vertical members taller than they would otherwise be. The presence of suspended floors can be turned to advantage in providing structural stability to the building by acting as a horizontal diaphragm (see Ch. 11).

An absence of internal divisions in a building of a significant floor area makes a framed structure an almost automatic choice. The roof, or upper floors, will generate loads which can only be satisfactorily handled by suitable steel or concrete columns around the edge, and the roof or floor structure itself will probably also require some system of framing to enable it to span from one side of the enclosed space to the other.

The presence of internal walls can contribute to the solution of how to support the upper structure by providing load-bearing facilities within the area enclosed and by the stiffening effect of the internal walls on the building generally.

In some commercial buildings, the imposed loads on the floors can be quite high – paper in bulk is heavy and, therefore, filing rooms, libraries etc. have a high floor loading. In most industrial buildings, the loads to be carried are usually quite large. But in all cases the loading stress to be allowed is greater than for domestic work. Typical recommended floor-loading allowances are shown in Table 10.1.

Table 10.1

Building use	Distributed load (kN/m²)
Dwellings	1.5
Offices:	
generally	2.5
with computing, data-processing equipment etc.	3.5
filing and storage	5.0
Shop display areas	4.0
Banking halls	3.0
Schools and colleges:	
assembly areas	4.0
classrooms	3.0
Hotels:	
generally	2.0
bars and vestibules	5.0
Churches	3.0
Factories	5.0 to 10.0

Much of the design consideration arising from the subject of imposed loads will be concerned with the internal arrangements of the building and an endeavour to site the heavily-loaded areas at the ground-floor level.

Once this has been achieved, there may not be any other accommodation requiring an upper-storey provision and thus the building is single storey. This is the reason why many factories have single-storied manufacturing areas but multi-storied office areas.

Having arranged the building layout, the designers must select a structural system capable of carrying the loads. If, for instance, the design team is forced into putting, say, an office computing room on an upper storey, for reasons of desirable internal plan organisation, then the structural engineer would probably choose a framed structure which can be adjusted in arrangement and size of members to take the local higher loading. If, on the other hand, the project is for a suite of small offices of normal use of limited height, a load-bearing wall system may be the best structural form.

When deciding the form of construction for the superstructure of a building, the designer must take into account the materials which could be brought to site. Two factors control this, local availability and site access. In this country, steel and concrete are equally available but, where the site is in a remote region, the cost of transporting the larger mass of the concrete (either in ready-mixed form or precast units) could well make it uneconomic in comparison to steel. But this cost advantage of steel could be overtaken by a still lighter structure of, say, aluminium tube, where this is appropriate for the building use. Where the distance of the haulage is a less important factor, the ease with which vehicles can reach the site may influence the choice. Sites in built-up areas must be approached by roads which can be narrow and contain sharp, constricted bends or junctions. In these circumstances, the movement of long lengths of steel or precast concrete may be a physical impossibility. The designer must, therefore, resort to a structural form based on small units such as bricks, blocks or precast concrete modules (for a pre-stressed post-tensioned frame) which can be carried in small loads, or else he must select an in-situ concrete system using either ready-mixed or, if the approach roads cannot take the weight or size of a truck mixer, site-mixed concrete.

Many systems of structure impart a characteristic appearance to the building. For example, the lightweight steel-trussed arrangement traditionally used for factory workshops produces the familiar saw-toothed building, whereas the heavier steel beam and stanchion system used as a framework for offices generates the box-like appearance many of these buildings exhibit. It is important, therefore, that the desired aesthetic appearance of the building is taken into account when choosing the form of structure, otherwise design work may proceed on a structural system which is perfectly suited to the building's function but which adversely affects the preferred elevational treatment.

The final controlling factor in the process of selection of a structural system is the one that influences all design decisions–cost. It is not just the cost of the system which must be examined but also the consequential costs. A steel frame may be cheaper than a concrete frame but the steel must be provided with a protection against fire, which the concrete does not need, and which can reverse the cost advantage. A concrete portal frame, for instance, exposed within a building requires a finishing treatment which may add more than the difference between the concrete and a more expensive laminated timber frame. Any framed structure must be clad to achieve the building enclosure, but a load-bearing brick structure both supports and encloses and requires no additional cladding. The same principle extends to the subdivision of the interior and details of finishes and services; the structural system affecting such things as the accommodation of service runs or ducts, lighting layouts (which get involved with down-stand beams), heating arrangements etc., all of which must receive attention if the final choice of superstructure is to be the most economic.

10.2 Structural forms

There are many ways in which the substance of the superstructure can be disposed to achieve the functions of support and enclosure, but most can be classed as either load-bearing wall systems in which the structural members are of a linear character, and may, or may not, also enclose the space; or framed systems in which the structural members are isolated supports carrying horizontal floor and roof elements, and a space-enclosing cladding (see Fig. 10.1).

10.2.1   Load-bearing walls

The load-bearing wall systems are either of the cellular type, which represents the majority of our historic buildings, or the cross-wall type. Either of these find application in industrial and commercial work, but since in this system the spans must be kept small and the building height restricted for practical and economic reasons, the range of suitable building uses is limited.

The cross-wall system of construction was shown in Chapter 1 to have applications for flats because of the small spans and repeated plan arrangements. It can also find a use in office and light industrial buildings, particularly the type built for small firms on industrial estates. The system consists of a row of load-bearing walls built at right angles to the length of the block, which provide separation between the occupied spaces and support for the floors (usually of concrete) and the roof. The front and back walls are completed with a lightweight infill panel or, in the case of workshops, it may be a large industrial door (see Fig. 10.1).

10.2.2   Portal frames

In many industrial estates, the small, light industrial units are restricted to single-storey buildings because of problems of floor loads and goods access to upper storeys.

Where this policy is followed, the portal frame system of construction is used. A portal frame is a structure with a rigid joint between the frame rafters or roof beams and the columns. This rigidity transmits a bending force to the columns which, by their resistance to this bending, stiffen the rafters or beams (see Fig. 10.2). The frame can be of either precast reinforced concrete, steel or laminated timber. The building enclosure is completed by purlins across the rafters supporting a corrugated industrial roof and brick or block infilling walls between the columns.

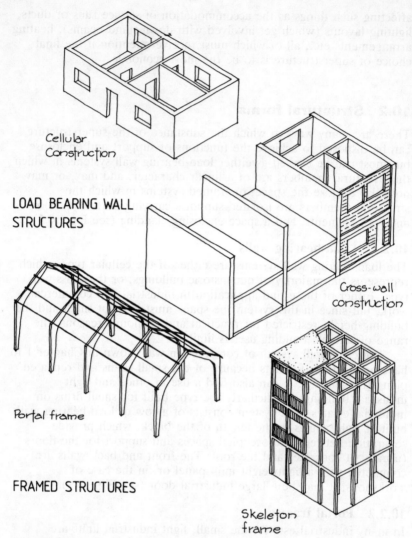

Cellular
Construction

LOAD BEARING WALL
STRUCTURES

Cross-wall
construction

Portal frame

FRAMED STRUCTURES

Skeleton
frame

Fig. 10.1 Forms of construction

The attractions of the system are that it is simple to fabricate
and erect, it produces a large uncluttered floor area and there is no
framework at roof level which might restrict the use of the internal
height. By the nature of the system, concentrated localised stresses
occur and, therefore, the imposed loads can only be light. For this
reason, portal frames are only used in single-storey buildings or as
the top storey of a multi-storey structure.

Special consideration must be given to foundation design

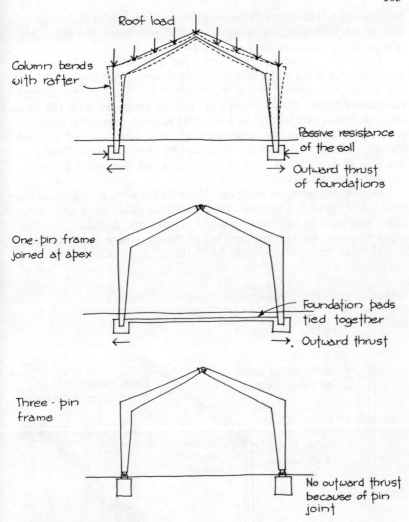

Fig. 10.2 Portal frames

because of a rotational force which can develop at the foot of the column; this is dealt with in section 10.3.

Portal frames may be rigid, one-pin or three-pin frames (see Fig. 10.2) depending, to a great extent, upon the size of the intended structure. Most manufacturers produce portal frames in components to be assembled on site, but with very small buildings the whole frame could be made up in one piece (provided it can be transported to site) in which case a rigid frame would be used. The one- and three-pin frames permit the frame to be made in two parts which are connected at the pin joints. Larger frames may be further

broken down into rafter or beam units and column units, which
have a specially designed jointing method to achieve the necessary
continuity of structure between the two.

Concrete and steel portal frames are available in all forms and
are widely used for industrial purposes as well as finding a ready
application in farming, where they provide the ideal barn structure.
Laminated timber portal frames are usually produced with the beam
and column as one unit. The basic unit is more expensive than
concrete or steel but, in buildings where a higher standard of finish
and appearance is demanded, they can be competitive in their
installed cost, since the cost of finishing is less, and offer a very
attractive appearance.

Details of concrete, steel and laminated timber portal frames are
shown in Fig. 10.3. Concrete frames are cast either as a one-piece
rafter and column unit to be jointed at apex and foundation, or as
separate parts, for bolting together as shown, where the frame size
would make transport difficult. Steel frames are made up from

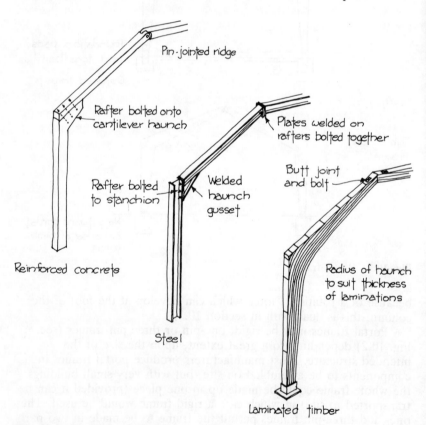

Fig. 10.3 Portal frame details

either a rolled steel section for both rafter and column, connected
by a haunch gusset at the knee joint, or a light, lattice-work
structure of angle- or tube-section members. The lamination of
timber frames is achieved by glueing together layers of thin timber,
bent to the required shape, and holding them in position in a jig
until the adhesive has set. When released, the timbers retain their
bent shape which is extremely strong and, when cleaned up and
varnished, require no further treatment.

10.2.3　Framed multi-storey structures

When more than one storey is required, the need is for a structure
which will afford flat floors and will carry the imposed loads. The
usual answer to this demand is a skeleton frame as shown in
Fig. 10.1.

In a skeleton frame, the upright load-bearing members (posts in
timber, columns in concrete and stanchions in steel) are arranged at
the corners of square or rectangular bays and are linked by a system
of load-bearing beams and non-load-bearing ties. The beams support
the floors and roof, the ties retain the upright members in position
at right angles to the beams and the whole structure is enclosed by
cladding attached to either uprights or the edges of the floors.

The system is very flexible and can be used, in one form or
another, for almost any building size or shape and all the variations
in imposed floor loadings normally encountered. Since much of the
frame and the other elements such as floors, roof and cladding will
probably be prefabricated away from the site, a regularity of
building form and standardisation of components is reflected in a
more economical use of resources. But this does not mean that a
skeleton frame cannot be used for buildings filling odd-shaped sites;
they just work out more expensive per square metre than a simple
rectangular building.

Many differing arrangements of uprights and beams have been
devised to meet the building user's requirements. Figure 10.4A
shows a typical frame based on rectangular bays in which the load-
bearing beams are placed along the long edges of the rectangle and
floor beams span the shorter distance. This is suitable where the
building has a general use, any internal divisions are light partitions
and there is no objection to internal, regularly-spaced uprights.

Figure 10.4B shows a modification of this arrangement where
the building plan has a central corridor with rooms of uniform width
each side. In this case the internal uprights are moved towards each
other to coincide with the fixed partition between the corridor and
the rooms.

In both Fig. 10.4A and B, the floor spans between two parallel
rows of beams and the ties at right angles to the beams carry no
load. Where heavy loads must be sustained, the ties are increased in
size to a load-bearing capacity and the floor is designed to be

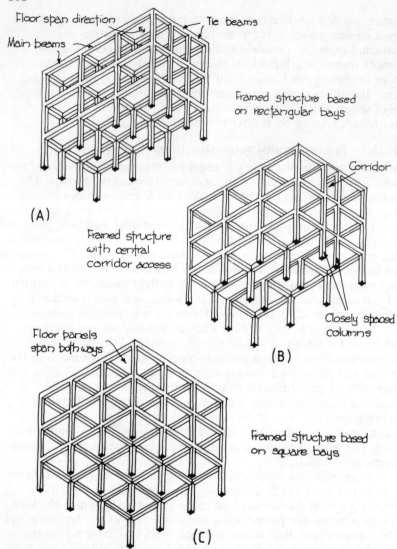

Floor span direction — Tie beams

Main beams

Framed structure based on rectangular bays

(A)

Framed structure with central corridor access

Corridor

Closely spaced columns

(B)

Floor panels span both ways

Framed structure based on square bays

(C)

Fig. 10.4 Framed structures

supported by all four perimeter beams (see Fig. 10.4C). To equalise the bending stresses and beam loadings, the uprights are usually positioned on the corners of a square bay.

10.3 Choice of foundation

As was explained in Chapter 1, the precise form of foundation for any building is decided by the superstructure standing on it, the building loads to be carried, the bearing capacity of the subsoil and the topography of the site. Building loads and site characteristics vary from one contract to another, but the way the form of superstructure influences the choice of foundations remains constant.

Where the loads are brought down to the ground along a line, as in the case of a load-bearing wall, the first choice for foundation is a simple strip of concrete, as for domestic construction. In many industrial and commercial buildings, however, the superstructure is framed and the loads arrive at the foundation level at isolated points. The simple solution to the problem of transferring these point loads to the subsoil is similar to strip foundations but is a square of concrete instead of a long strip. It must be a square since the load disperses uniformly in all directions through the concrete and its size is controlled by the magnitude of the load and the bearing capacity of the subsoil.

Where heavy loads, poor soil or a combination of both occur, the foundation designer must seek other means of supporting the superstructure, and often turns to a piled foundation. Foundation piles are columns of concrete driven or formed in the ground in such a way that the loads are carried deep into the subsoil to levels where a strong stratum exists. For lineal loads, these piles are linked by a reinforced concrete ground beam just below ground-level on which the wall stands. For point loads, the caps of a group of piles are connected by a reinforced concrete slab on which stands the structural upright.

Where a portal frame system of construction is used, the bending force in the column induces a rotational effect and outward thrust at its foot. This is transferred to the foundation, which must be designed to withstand it. Two ways of dealing with this are shown in Fig. 10.2. In the first, the outward thrust is retained by the passive resistance of the earth and in the second the bases are linked by a concrete or steel tie beam below the floor. If a pivot is introduced between the column foot and the foundation pad, as in the case of a three-pin frame, none of the rotational force is transferred to the foundation and hence there is no outward thrust.

10.4 Provision of natural light

Most multi-storey buildings are small enough in plan for the windows in the perimeter to provide natural light to all the interior, but many single-storey industrial buildings cover such a large area that normal windows are of little use and natural light can only be obtained from windows in the roof. Despite the fact that present policies are turning away from natural light as the source of illumination, preferring to rely on artificial light which costs less than the heat loss through the glazing, the provision of some natural lighting, in case of a power cut, is a wise precaution. Windows are also necessary to avoid a feeling of claustrophobia.

Windows in the enclosing cladding of a framed structure present very little problem and, in some cases, it is merely a matter of using panels of glass rather than the opaque material used elsewhere in the envelope. Roof glazing can be either a translucent sheet substituted for the normal roof sheet, a glazed roof light set into the roof covering or glazing fitted into a modified roof shape. Two such modified shapes are a North light roof and a monitor light roof (see Fig. 10.5). The North light roof, producing the familiar factory outline, is intended to set the glazing at a steeper angle than the roofing sheets so that it is less likely to leak. It has been widely used in the past but, for modern standards of energy conservation, the area of glass is too great and some must be replaced by insulated roofing. The monitor roof system has a smaller area of glass, which is also pitched at a steep angle to prevent leaks, and produces a flat ceiling inside the building.

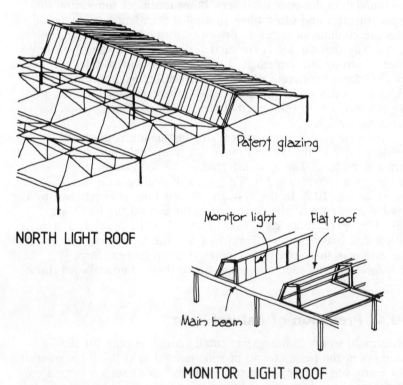

Fig. 10.5 Roof glazing

Chapter 11

Stability and cladding of framed structures

11.1 Stability of frames

Most structural frames are required to be rectangular in form, because the components to be fitted to and accommodated within them are also rectangular. The geometrical properties of a rectangle suffer from a serious disadvantage: the shape can be altered without changing the lengths of the sides. All plane shapes, except one, exhibit this characteristic, the one exception being the triangle. Therefore, a building composed of rectangles will distort if subjected to appropriate forces whereas one of triangles will not but, unfortunately, most building users require the enclosing walls to be upright with a flat, level floor and ceiling.

The forces which induce distortion in rectangular frames are any which are not coincidental with the axis of the upright members. The most severe, non-coincidental force is wind load. Wind acts in varying ways but always at, or nearly at, right angles to the axis of the columns. On the windward side of the building, the effect is a positive pressure trying to push the obstructing building over; on the leeward side it is a suction or negative pressure; and at the roof level it may be either, depending on the roof shape (see Fig. 11.1).

Less severe than wind loads but nonetheless forces which could cause distortion are those arising from eccentricity of loading. Large cantilevered structures or pieces of heavy equipment attached to the framing import bending forces to the upright members and, if these are insufficiently stiff or inadequately braced, the complete structural

112

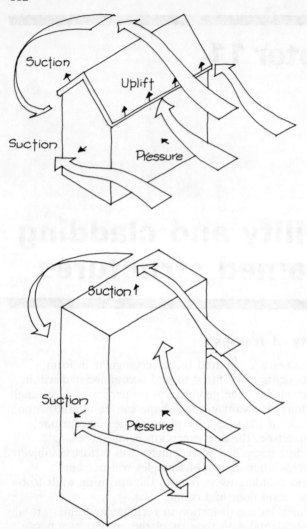

Fig. 11.1 Wind around buildings

frame can be distorted, leading to the displacement of external cladding and fractures of internal floors and walls.

11.2 Methods of lateral restraint

As explained above, the triangle is the only shape which cannot be altered without altering the length of the sides; one method of obtaining lateral restraint is to use the desired rectangular frames but to break them down into triangles by the use of a diagonal brace. This technique is very common in many types of framed structures – bridges, ships aircraft etc., as well as buildings.

Although it is possible to distort a rectangle without altering its sides, the angles must change. It ceases to be a rectangle and becomes a rhomboid in which the angles are no longer 90°. If this change is prevented from happening, the rectangle cannot distort and the building remains stable. This principle finds application in concrete structures where rigid joints are readily formed.

Another way of preventing distortion is to make use of a diaphragm. This principle is demonstrated by a sheet of paper which can be folded or bent with ease, but cannot be changed from the rectangular shape in which it was cut. By attaching the paper to a framework which prevents it being folded or bent, a rigid structure is formed. Sheets of paper are not strong enough for building loads, but sheets of concrete or panels of blocks are capable of sustaining the forces encountered. Plywood is also used in this way, as explained in Chapter 5.

All the preceding methods have been devised to make the frame itself rigid. An alternative approach is to attach the unstable rectangular structural frame to a solid object, thereby preventing it from moving. The solid object usually adopted for this purpose is a mass core structure within the frame, generally containing lifts, staircases, toilets etc.

11.2.1 Triangulation of frames

This, the simplest method of stabilising a frame, is usually applied to steel and timber framework, seldom to concrete. It is the principle underlying the design of all roof trusses and trussed rafters and, in some steel structures, has been exposed as a feature of the building design.

Figure 5.2 shows examples of the triangulation of roof frames in timber which follow the same lines as similar frames in steel. The 'W' or Fink trussed rafter shown is one of the triangulation arrangements used to break down the main triangle of a roof frame into smaller triangles in which the stresses in the individual members are more easily handled. Other truss forms are shown in Fig. 11.2.

More typical of the principle of bracing a rectangular frame by triangulation is the trussed purlin shown in Fig. 5.2 in which there are horizontal top and bottom members connected by vertical members, and the resulting rectangles are retained in their shape by a diagonal brace, producing the typical 'N' formation. This is the most common arrangement and is also used in the construction of steel lattice girders. A girder is a horizontal structural member made up from various steel sections, welded, riveted or bolted together. A steel beam is also a horizontal structural member but it is formed by

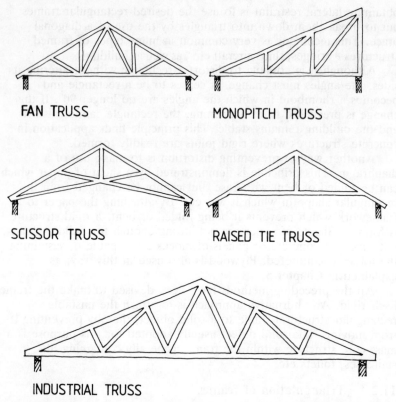

FAN TRUSS   MONOPITCH TRUSS

SCISSOR TRUSS   RAISED TIE TRUSS

INDUSTRIAL TRUSS

Fig. 11.2 Truss formations

rolling the steel, either hot or cold, into the desired shape. Girders can be made much deeper and stronger than beams. They are used in some buildings to achieve large clear floor areas and, in particular, are used in bridge construction. Plate girders are made of a plate of steel on edge, forming the web of the beam, with flanges attached along the top and bottom edges of strips of steel plate welded on, or lengths of steel angle riveted in position. The lattice or trussed girder form, is much lighter than a plate girder and more economical in the use of steel.

The triangulation of building frames to provide structural stability follows a slightly different pattern to that used for trusses, purlins and girders. In a steel-framed multi-storey structure, not all of the rectangles formed between the beams and the stanchions need to be triangulated or braced. The usual practice is to select a number of pairs of stanchions, between which bracing can be fixed without inconvenience to the plan and connect the heads of each to the bases of each, with a diagonal tie, thus creating an 'X' arrangement. This is repeated on each floor in the same position to form rigid vertical panels within the building. The wind pressure on the building enclosure is transferred to these rigid panels through specially constructed floors, at selected levels, attached to the braced

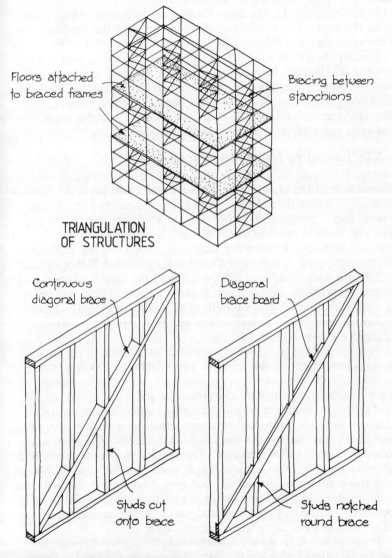

Floors attached to braced frames

Bracing between stanchions

TRIANGULATION OF STRUCTURES

Continuous diagonal brace

Diagonal brace board

Studs cut onto brace

Studs notched round brace

BRACING OF TIMBER FRAMES

Fig. 11.3 Triangulation of frames

frames. The 'X' arrangement is better for this application because, unlike a girder in which the direction of action of the load is always vertically downwards, with a building frame the direction of wind pressure load is variable and can completely reverse. By bracing across both diagonals, the members can both be designed as ties, which are easier to achieve, rather than struts, which must possess a minimum degree of stiffness (see Fig. 11.3).

In timber framing, the diagonal bracing intersects many, if not all, of the studs. The traditional method was to make the brace out of the same size of timber as the studs; the latter were splay cut to fit on to the brace (see Fig. 11.3). More recent developments have shown that notching the studs to receive a brace of approximately 25 × 100 mm timber has improved their ability to carry a load (by maintaining their continuity) and has been proved to achieve adequate lateral stability. The brace must be let into the studs to produce a flat surface to receive the subsequent facing.

11.2.2 Lateral restraint with diaphragms

The clearest example of the use of a diaphragm to restrain movement is in the cladding of timber frames with plywood. This is much easier to do than the diagonal bracing described in the last section, and is more efficient because the studs are not cut in any way. The method consists solely of nailing specially produced sheathing plywood to the outer face of the whole of the studwork. The plywood used is generally Douglas Fir plywood 2440 mm long × 1220 mm wide × 12.5 or 18.5 mm thick; the face veneers are of low grade and show many manufacturing defects. These defects are not such as will affect the strength of the diaphragm and, since the plywood is completely hidden in the completed structure, its appearance is not important.

By nailing these sheets of plywood across the stud faces, any sideways movement of the frame is eliminated. In addition, the plywood stiffens the studs, preventing them from bowing and thereby increasing their load-carrying capacity.

The same diaphragm effect is present in monolithic concrete buildings. When a reinforced concrete structure is cast in situ, the columns, beams, floors and any reinforced concrete walls are all formed as a continuous structure. The method is slow and relatively expensive but a number of advantages arise, not the least of which being the inherent rigidity of the building. This rigidity is partly due to the unalterable beam-to-column joints (which is dealt with in more detail in the next section) and partly to the diaphragm of the floors and walls stiffening the other members.

In situ concrete floors are inadequate in themselves to brace a complete building frame – some vertical element is required against which the floors can bear. Such vertical elements can be found in the walls surrounding the stair well and lift shafts or in the end walls. By casting these in concrete they can be linked to the floors which, in turn, are monolithic with the beams and columns, and the completed structure is very stable (see Fig. 11.4).

In some cases the stability achievable by a structure completely cast in situ is greater than required and does not warrant the additional cost. For these situations a compromise between cast in situ and precast techniques may well offer the most economic

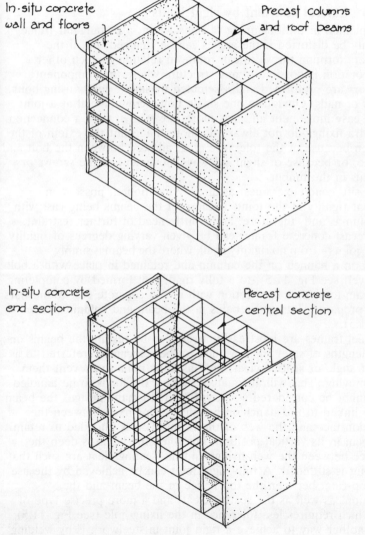

In·situ concrete wall and floors

Precast columns and roof beams

In·situ concrete end section

Precast concrete central section

Fig. 11.4 Part monolithic structures

solution. With such a compromise, the end bays of the building and possibly a central section may be cast in situ with these rigid monolithic vertical elements connected by a precast system of columns, beams and floors. The in-situ parts of the building would be the logical location for stairs and lifts which, because of their special requirements for openings in the floors, create difficulties in a precast system and, furthermore, the reinforced concrete walls which assist in the resisting of lateral forces also provide the fire-resisting enclosure required for fire-escape purposes.

11.2.3 Lateral restraint by rigid joints

As explained at the beginning of this section, a rectangular frame can only be distorted if the angles can be changed, i.e. if the members forming the sides can rotate in relation to each other. Such rotation is possible in any structure where the component members are preformed and assembled by simple joints using bolts, screws or nails. Obviously one single bolt, screw or nail at a joint allows easy movement and acts more like a pivot than a connection, but extra fixings do not always do more than limit the extent of the rotation because of the tolerance of fit allowed between the bolt and its hole, or because of slight sideways movement of the screws or the nails in the timber.

In-situ concrete frames, as already mentioned, possess an inherent rigidity at the joints because of the beams being cast with the columns and, therefore, are in little need of further restraint.

Precast concrete frames can be given varying degrees of rigidity at the joints – from no rigidity at all, where the beam is simply seated on a haunch on the column and retained in place with a bolt or dowel (see Fig. 11.5), to a fully rigid joint formed by completing the beam-to-column connection with in-situ concrete cast round bars set in prepared slots and rebates in the beam and column (see Fig. 11.5).

Steel frames are most easily assembled by resting the beams on short lengths of steel angle welded to the stanchions (referred to as a shelf angle or seating cleat) and inserting bolts to prevent them from moving. This will adequately support the loads to be handled but cannot be considered as a rigid joint. In many designs the beam is also linked to the stanchion by angle cleats fixed between the stanchion face and the web of the beam. This is intended to retain the beam in its vertical position, but when the beam is deep the distance between the web fixings and the seating cleat are such that the joint is stiffened. A fully rigid joint can be achieved by the use of a top cleat between the top flange of the beam and the stanchion, as well as the seating cleat, and a more precise type of bolt which requires less tolerance in the fixing hole (see Fig. 11.5).

Another way to achieve a rigid joint in steelwork is by welding the members together. On a building site it is a difficult operation

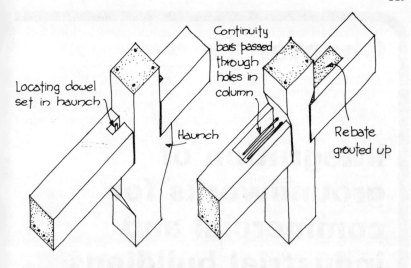

PRECAST CONCRETE STRUCTURES

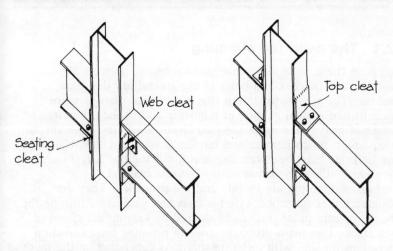

STEEL STRUCTURES

Fig. 11.5 Structural joints

to weld beams to columns in position in a multi-storey structure, but it can be achieved if the beam is first seated on a cleat and temporarily held to prevent it from moving. With the beam correctly positioned, the welder then runs a fillet weld all round the junction between the end of the beam and the stanchion face.

Chapter 12

Integration of groundworks for commercial and industrial buildings

12.1 The need for planning

Except in the case of very small excavations or inaccessible
positions, all trenches and holes in the ground are dug by a
machine. The reason for this is that mechanical plant is more
efficient than a team of men in removing and hauling the large
masses of heavy material usually involved. Machines, however, are
much larger, much heavier and far less flexible or adaptable than a
man and this must be taken into account if the cost advantages of
the greater efficiency of machinery is to be realised.

Most excavators are mobile and self-propelled. They are
mounted either on rubber-tyred wheels and steered by turning the
wheels, or caterpillar tracks and steered by varying the speed of
each track. Two main problems arise in planning the excavation
operations on site, with differing solutions depending on the method
of mounting. Firstly, the machine has to be transported to the site
and to and from its working position on the site. Secondly, its work
must be arranged so that it can carry out its manoeuvres without
either falling into completed excavations or endangering above-
ground site features. Movement along public roads is usually by
transporter but, in the case of tyred machines, they can be driven
on the road from one point to another if the distances are short
enough to make this economic, bearing in mind the slow speed of
the vehicles, the time it takes and the frustration caused to other
road users.

Site planning commences with the question of getting the excavator to its working position. This presents no difficulty if the site is an open field but if it is, say, a congested factory premises the width of the entrance, the space between buildings and any overhead obstructions such as pipes, wires or connecting bridges will dictate the size and type of equipment that can be used, in addition to other factors such as the type of work involved, the quantity of subsoil to be moved, the programme of working etc.

Having persuaded the equipment to its area of operations, the way in which the work is tackled must be carefully studied. When the excavations are complete, the site will be laced with trenches, perforated with holes and covered with heaps of subsoil, all of which restrict the movement of the machine which created these conditions. If excavation is started at the point of entry and continues across the site, the excavator will finish up marooned in a far corner until such time as the groundworks are completed and the trenches filled in again. This means leaving expensive equipment idle when it could be working on another site.

Another planning point which must be considered is the working area required for the selected plant. All machines described have a slewing circle which extends beyond their tracks and which must be accommodated clear of any buildings, posts or trees. As well as rotating, many machines are intended to travel across the site with excavated spoil. These paths must be planned to avoid any overhead obstructions and with possible damage to existing surfaces in mind. Not only must consideration be given to excavating equipment, the lorries required to transport away any surplus spoil and those which will be arriving loaded down with ready-mixed concrete, for the foundations, must also be thought about. A fully-loaded ready-mixed concrete lorry can weigh up to 24 tonnes and has a turning circle of 15 m, and yet must be driven across the site to a point of discharge within 3 m of where the concrete is required. Alternatively, and more expensively, the truck can be left on the road and the concrete transported in smaller batches by dumper truck or continuously by a concrete pump.

12.2 The building programme

Although more or less the same in their nature, the purpose of groundworks can be separated into two groups: those concerned with the building structure and those concerned with the services to the building and the surface of the site around the building.

Groundworks in connection with the building consist of site levelling and stripping; trenches for strip foundations; pits for pad foundations; bore holes for pile foundations; or excavations for basement structures. Building services and site works require trenches for pipes, tubes and cables; stripping and shaping to formation levels for paths, drives and roads; and scraping, transporting and grading for landscaping purposes.

The separation into two groups reflects the closeness of the link between the groundworks and the programme for the erection of the building. Obviously, the site preparations of levelling, stripping and foundation excavations must be the first item in the process of building, but the other group of operations is not closely tied to the timing of the building works, although it must be completed within the contract period. For convenience, much of the groundworks in connection with the services is left until late in the building programme; roads, drives and pavings may be an early priority if they are required to ease difficulties of access, or they may be left until the end of the programme to avoid damage; landscaping and planting is always the last of the items to be carried out.

Deferring service trenches until later in the programme reduces site congestion at the beginning of the building progress, making the transport of materials easier and avoiding problems of trying to erect scaffolding on disturbed ground. It is also to be recommended where the services, such as drain runs, may be damaged by heavy construction vehicles if they are installed at the start of the works.

Certain services works cannot, however, be left until later in the programme. Any supplies or facilities which must pass under or reach into the completed building must be installed, or provision made for their later installation, before or during the formation of the substructure. Generally the services involved in this manner are drainage and water supply. Electricity, gas and telephone services frequently terminate at a point in a purpose-made box, in the external wall of the building, or in a purpose-built meter room. Branch drains must be laid to pass through the foundation structure and below the basement or ground-floor slab to the points where they are required for connection to soil pipes, internal gullies or WC pans. The water main in a building must rise at a point not less than 600 mm from any external wall and, therefore, a suitable trench must be cut for this purpose. On many sites, a sleeve pipe is installed in the early stages through which the water pipe is threaded when the rest of the external service work is carried out. In both cases, the work must be carried out with accuracy to ensure that the correct terminal positions and external connection levels are achieved, and at the right point in the programme, before or during the construction of the foundation works and before the laying of the floor slab.

12.3 The effect of the site, soil type and existing services

In the majority of commercial and industrial developments, the need is for large floor areas, usually all in one plane. Some, but by no means all, of the building sites available for this type of development are conveniently flat and level over an area large enough to accommodate the building. With the rest of the sites, the slope on the ground must be allowed for in the design of the building, its foundations and its services, including considerations of the way these works are to be carried out.

Many are the solutions available to the designer and the builder to the problems set by sloping sites, but these are beyond the scope of this chapter. However, the choices made will have an influence on the groundworks required and the integration of the various operatives required. One solution often adopted in these circumstances is to level the area to be built upon by a 'cut-and-fill' (see Fig. 12.1). In this, part of the site is removed at the high side of the building area and deposited on the low side, producing a level platform. Where this is chosen, the cut-and-fill operation must precede any other groundworks and the filled area of the site must be thoroughly compacted to support the heavy equipment running over it in the next stage. Having levelled the site with a tractor shovel or scraper, the excavators and trenchers can then move in to take out the foundation and service trenches.

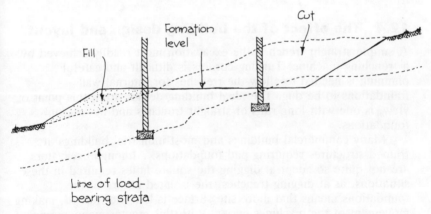

Fig. 12.1 Cut-and-fill

Not only must care be taken over the capacity of the filled area of a site to carry the weight of large earth-moving plant, as mentioned in the last paragraph, due consideration must be given to the natural ground surface as well. Sites with soft upper layers of soil or a high water table may call for special equipment and particular planning of the work. One of the problems to be faced is that with loose soils there is a tendency for trench walls to collapse. This tendency is heightened by placing a surcharge on the ground surface – such as occurs when a heavy vehicle passes. In these circumstances the layout and sequence of excavation of the trenches must be studied to avoid the equipment causing the collapse of one trench while excavating another. Any such trenches in which men will be working must be supported but, say, a narrow trench to take just a water pipe would probably not be so supported. However, carelessly-driven machines could bring down the trench wall, making it necessary to re-excavate before the pipe can be laid. In extreme cases, it may be necessary for the excavator to make several passes before the full depth of trench is achieved to allow timbering to be installed at each stage, otherwise the weight of the machine could collapse the trench just excavated.

Much of our building land has been the site of buildings before and contains hidden foundations and unrecorded service runs. There is no way the builder can make allowance for these beyond keeping a vigilance during the progress of the work. Where existing buried services are known to exist, their presence can have the effect of reversing the usual order of foundations first, followed by the service installations. It could mean that the first groundworks operation is to excavate to expose these buried services, to allow them to be diverted or strengthened before any building work can be put into effect.

12.4 The effect of the building design and layout

A simple straight trench is the excavation most readily achieved by a trenching machine. Turning corners is difficult and careful planning is needed to allow the trenches for internal wall foundations to be dug. The ideal building design, from this point of view, is one with long runs of straight trenches and no internal foundations.

Many commercial buildings and most industrial buildings are framed structures requiring pad foundations. Although excavators are not quite so adept at digging the square holes required in these situations, as at digging trenches, the isolated nature of the foundations means that more site surface is left undisturbed, making movement of the machines easier. With this greater manoeuvring ability, it often becomes possible to excavate the service trenches at the same time, thus saving a second visit to the site by the excavating equipment.

Another situation calling for the service trenches to be taken out at the same time as, or even before, the foundations arises from the site layout where, say, a drain run occurs parallel to, and near, a wall of the building. Most trenching machines straddle the line of the trench and thus require a clear working space each side. If the distance between the drain and the wall is less than the working space required, the drain must be laid first and the foundations and wall built later, otherwise the wall would prevent the excavation of the drain trench. Since the wheels or tracks of the machine will pass along or very near to the drain trench when excavating for the foundations, it would probably be necessary to lay steel plates on the ground to distribute the load.

12.5 The effect of foundation type

Foundation structures can be grouped into two types, pile foundations and spread foundations. The former seek to place the building loads deep into the subsoil and the latter distribute them near to the underside of the building. Most other groundworks take place fairly near the surface and, therefore, the spread foundation systems are more likely to affect other work on site than the pile foundations. However, pile-boring or pile-driving equipment can be large and heavy and other work must be considered when planning the building programme.

The integration of foundation works with other groundworks may be further influenced by the mechanical plant used. If spread foundations are used (concrete strips, pads or rafts), the excavator needed to form the trenches or holes could also be the machine used to cut the trenches for drains and services and to strip and shape the surface for roads, drives, paths etc. In these circumstances, the economics of planning the use of heavy plant would suggest that, once on site, the machinery should complete all the earth moving, rather than making two or more visits, thereby increasing transportation costs. As mentioned earlier in this chapter, there may be other reasons why it may not be practical to complete all the groundworks in one operation, but if these do not apply, then spread foundations and other work below the surface would be programmed together.

Surface disturbance by pile foundations is only slight compared to spread foundations. The site is not cut up by trenches, the spoil heaps from bored piles are only small and isolated, and driven piles produce no spoil at all. The equipment is often mounted on a tracked vehicle, sometimes a small mobile crane is modified for the purpose, and can manoeuvre around a site reasonably easily provided that no other works have been carried out. Since the pile foundation equipment is quite different to that required for other groundworks and pile foundations do not present many difficulties for following site works, the two are generally separated in the building programme, the foundation installation being completed, usually under a separate sub-contract, as one of the first operations followed, when convenient, by the other groundworks.

12.6 The effect of site layout

Every building contract has features which make it different to all
the others, and the planning and integration of groundworks reflects
these differences. The two extremes which will be dealt with in this
section are probably, on the one hand, a relatively small extension
to, say, an industrial complex, comprising two or three bays added
on to the end of an existing workshop and, on the other hand, the
development of a complete industrial estate.

In the first instance, there will be a strong influence exerted by
the existing buildings, their established services and the pattern of
use applied to them. The work, particularly the physically disturbing
and messy groundworks, must be arranged to suit the convenience
of the occupants and to permit production to continue
uninterrupted. Irrespective of the constraints of construction
requirements, the excavations must be timed, for instance, to
coincide with a period when large earth-moving lorries can enter
and leave the site without causing trouble; any service installations
must be so arranged that the inevitable interruption of existing
supplies to allow a connection to be made occurs when those
supplies are not required; roadworks connecting with, or altering,
existing pavements may have to be programmed in the same way as
the services. Beyond these points, the integration of the groundwork
is subject to the considerations already discussed in this chapter.

A new industrial estate or any other complex of separated
buildings presents a totally different set of controlling factors. It is
usual for this type of work to take place on undeveloped land and,
therefore, is not subject to any of the constraints outlined in the last
paragraph. One of the first concerns of the builder is surface
transportation of goods, and involves the establishment of an access
point and suitable surfaces along which vehicles can pass no matter
what the weather. If, as is normal, the estate or complex is served
by a pattern of roads, the groundworks for these will be started
before the buildings, and a part of the road pavement structure laid
(the final surfacing is deferred until the end of the contract to avoid
the possibility of it being damaged by the construction traffic).
These preliminary roadworks would be linked to the existing road
serving the site at the point where the final connection will be
made, thus providing a satisfactory access. Laying the road
foundation structure at the start of a contract not only provides
means of on-site movement, but also ensures that the base on which
the finished road pavement will be laid is thoroughly consolidated.

Prior to laying the road base, it is necessary to carry out any
work below the line of the road. This usually involves laying the
main sewer runs, including any branches serving the buildings on the
opposite side of the road, from the mains drainage run. Since
sanitary facilities are required for everybody working on-site, it is

quite usual to connect this main drain to the existing sewer at this stage to provide a means of disposal.

Provision must also be made for the service runs which cross the road. It is not common practice to install water, gas, electricity and telephone services over the whole site at the start of a contract. Water, electricity and telephone must be run to the site offices and accommodation, but could be damaged if laid over the whole site prior to the main earth-moving jobs of foundation excavations and landscaping. To allow service runs to be installed without disturbing the road structure, ducts, usually of pipe, are laid at the appropriate depth from one side of the road to the other for each of the service mains opposite each of the proposed buildings.

Having cleared the site, installed the main drain runs and service ducts, and laid the road foundations, the builder can then start on the groundworks for the substructure of the buildings and such landscaping of the site as is required. Whether the external drainage branches and service runs are put in at this time, or left until later in the contract, will be determined by the factors explained in sections 12.3 and 12.4.

Chapter 13

Superstructure and cladding

13.1 Programming framed structures

The choice of a frame to carry the structural loads of a building not only affects the design and appearance of that building, but also exerts a strong influence on the programme of construction. The process of erection of framework is much faster than that of building a load-bearing wall. Thus, the roof supports are in place much sooner with a frame, but the building has not been enclosed. In a brick building it takes longer to reach roof level but, when this is achieved, the building has also been provided with an enclosing envelope.

There is an advantage in getting the roof on early: it provides protection to the rest of the work – and the workmen. This does not mean that, with tall buildings, all enclosing cladding of the frame is left until the roof is on, but with low-rise framed structures it is not uncommon for the roof structure to follow the frame erection immediately and the wall cladding to follow that.

The programme for constructing a framed building is firstly to cast the foundations, checking carefully their position and level, and the location of the holding-down bolts; secondly, to erect the frame and then to fix either the roof covering or the external cladding and, finally, to complete the enclosing envelope.

13.2 The erection of the frame

The first operation in the sequence is, as mentioned above, the casting of the foundations. When the bases are intended to support a steel frame, holding-down bolts are cast into the concrete of the base, ready to be passed through holes in the steel base of the column. If it is to be a reinforced concrete framed structure, the foundation will be formed with either pockets to receive the bottom end of the column or holding-down bolts which will locate with holes in a steel plate welded to the reinforcement (see Fig. 13.1).

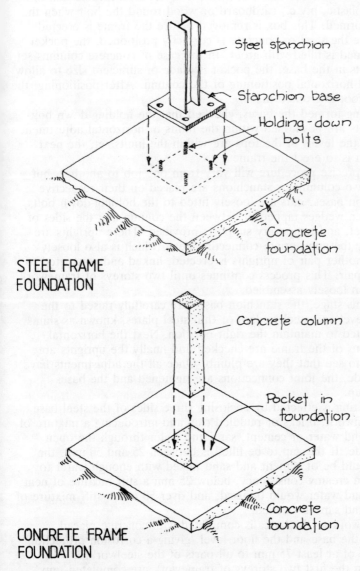

STEEL FRAME
FOUNDATION

CONCRETE FRAME
FOUNDATION

Fig. 13.1 Frame foundations

The holding-down bolt method makes access to the column base easier for levelling and fixing purposes.

The tolerances in framed structures are very small and due allowance must be made for the lack of matching accuracy on site. This, as far as holding-down bolts are concerned, is achieved by installing them in such a way that they have a limited amount of horizontal movement at the beginning of the erection process. The usual way to provide this movement is to set a removable box of foamed plastic, p.v.c., cardboard or wood round the bolt when the base is formed. This box is removed before the frame is erected and, once the steel or concrete is correctly positioned, the pocket thus formed is filled with grout. In the case of concrete columns set in pockets in the base, the pocket is made of sufficient size to allow a limited horizontal positioning of the column. After positioning, the column foot is grouted in.

Having formed the bases, checked that the holding-down bolts or pockets are positioned within the limits of horizontal adjustment and that the levels of the tops are within the limits set, the next operation is to erect the frame.

The precise procedure will vary from one job to another, but usually two columns or stanchions are placed on their respective foundation bases, nuts are loosely fitted to the holding-down bolts, or wooden wedges tapped in between the column and the sides of the pocket, and temporary supports provided. These uprights are linked by the appropriate connecting beam which is also loosely fixed. Another pair of uprights is erected, linked and connected to the first pair. This process continues until two storeys of framework have been loosely assembled.

At this stage, the stanchion bases are carefully raised to the correct level and steel wedges or thin steel plates, known as shims, are inserted to maintain the right position. Next the horizontal dimensions of the frame are checked and finally the uprights are checked to see that they are plumb. Once all the adjustments have been made, the joint connections are tightened and the bases grouted up.

Grouting is effected by enclosing three sides of the steel base with formwork, bricks or puddled clay and introducing a mixture of cement and water or cement, sand and water through the open fourth side. If the gap to be filled is between 25 and 50 mm, the grout would be of cement and sand mixed with enough water to produce a creamy consistency, below 25 mm a stiff mixture of neat cement and water would be used, and over 50 mm a stiff mixture of cement and sand is needed.

The work to the bases is completed with concrete placed between the base and the floor level, giving a corrosion protection thickness of at least 75 mm to all parts of the steelwork.

When the first two storeys of framework are completed, any

subsequent storeys can be added, checked for plumbness and secured.

13.3 Cladding

Completion of the building envelope is achieved by cladding the framework with one of a variety of products. Whichever one is selected, its construction must be capable of fulfilling all the functions of the external envelope, except that of structural support, and the functions peculiar to cladding systems. These functions are:

(a) the panels must be capable of spanning between the framing members to which they are attached;
(b) they must be able to carry positive wind load across this span;
(c) they must not be dislodged or distorted by negative wind load;
(d) they must resist all weather conditions;
(e) they must conserve energy by insulation and draught resistance;
(f) they must achieve an adequate fire resistance for the location in which they are fixed;
(g) they must possess whatever standard of sound insulation is specified;
(h) they must be the right size for economic transportation, handling and fixing.

13.4 Cladding panels

The types of products discussed in this section are all designed as claddings, i.e. they enclose the structural framework as a sheath, leaving none of it exposed. The alternative to this is infill panels which fit between and show the structural frame members.

Since framed structures are no use as buildings without an enclosing cladding, and claddings are not required unless the building is a framed structure, the development of both stems from the same period in time – the 1940s and 1950s – although the first metal skeleton framed building was much earlier – the Menier chocolate factory built near Paris in 1871/72, which was followed by the Chicago skyscrapers of 1890–1900.

One of the two original materials to be used for cladding panels was the latest development of that time, reinforced concrete, and the other was curtain walling derived from the design of greenhouses. Later developments have led to glass-reinforced cement, glass-reinforced polyester, profiled cladding and sheet metal panels.

13.4.1 Concrete cladding panels

Concrete is a dense material and so, for easier handling, cladding

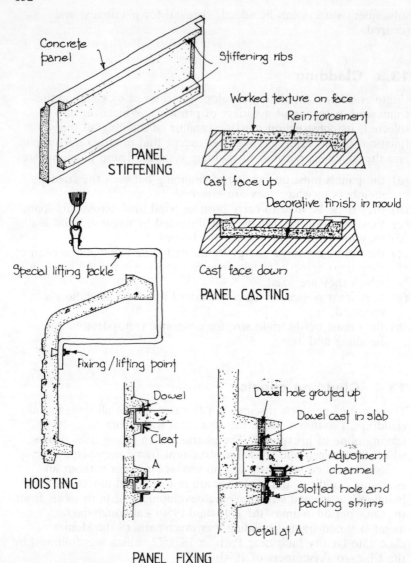

Fig. 13.2 Concrete cladding panels

panels are kept as thin as 50 mm, with the edges of the panels thickened or worked into fixing ribs to stiffen them (see Fig. 13.2).

A wide range of finishes can be worked into the surface either by modelling or texturing or by making use of the colours of the concrete aggregate generally, or a specially selected stone aggregate introduced near the surface. Panels can be cast either face up or face down. In the face-up position, the panel face can be worked to a limitless variety of finishes, each slightly or greatly different, as required. If cast face down, the mould can be lined with either a shaped panel to produce a regular repeated geometrical pattern or a layer of flints, marble chips or other decorative stone which becomes part of the concrete when the panel is cast and is exposed by washing, brushing or sand blasting when the panel has set (see Fig. 13.2).

The panel size must be large enough to span between supports but small enough and light enough to transport by road and handle on-site. The most economical width for panels to be transported flat is between 1200 and 3000 mm; they can be carried on edge or in pairs supported by an A-frame structure: in either case the maximum height when loaded should not exceed 4880 mm. To allow panels to be lifted and accurately positioned, it is recommended that the weight should not exceed 7.0 t.

Hoisting is done by means of special lifting points cast into the panel during manufacture and possibly specially designed tackle for profiled units (see Fig. 13.2). Fixing is by means of angle cleats or dowels and must allow for adjustment at the fixing position (see Fig. 13.2).

Accurately made and carefully installed, concrete panels are a very durable cladding, but their weathering properties are variable and may give rise to an unacceptable appearance. Failure is seldom due to the panel itself. It is due to either bad design, e.g. not allowing enough thermal expansion movement, or bad installation, e.g. inaccurate positioning of fastenings reducing their efficiency..

13.4.2 Glass-Reinforced Cement (GRC) panels

These panels are composed of a mixture of ordinary Portland cement, sand and water to which is added alkali-resistant glass fibres. Although the proportion of glass fibre is low – 3.5 to 5 per cent by weight – the resulting material is very tough and, because of the absence of corrosion in the reinforcement, can be used in panels as thin as 10 mm.

There are two methods of producing the panels, premix or spray. In the premix method, the constituents are mixed to a paste and the panel formed by casting, press moulding or slip forming. The spray method uses a wet mix of mortar and a separate supply of chopped glass fibres. Each is simultaneously sprayed from dual-spray head into the required mould. The method maintains better control over fibre orientation (the factor which determines the strength of GRC) than the premix method and is generally preferred.

Being a moulded product, the surface profile or texture is only limited by considerations such as the economics of re-usable moulds but, since the mixture is predominantly cement, it dries to a dull grey uniform colour with no coarse aggregate to provide interest. To improve the appearance, pigmented colourings can be added to the mix, specially prepared masonry paint can be applied or the GRC can be sprayed on to a p.v.c. film.

The thermal insulation value of GRC is low–a 20 mm thick flat panel has a U value of 5.0 W/m²K – and, although non-combustible and not easily ignited, a single skin of GRC will not maintain a satisfactory integrity when exposed to fire. For these reasons, GRC is generally used as a sandwich panel. If two 10 mm sheets of GRC are combined with a 50 mm core of polystyrene bead aggregate and cement, the resulting 70 mm panel can be shown to have a U value of 2 W/m²K and a fire resistance of 2 hours; it is also capable of spanning three times as far as a single skin of GRC. The thermal insulation value is still inadequate and needs an inner lining added to achieve current standards but is a great improvement on 5 W/m²K.

As with all cement-based components, GRC suffers an initial drying shrinkage and is prone to movement in use, due to moisture changes. A disadvantage of the material is that these movements are larger than for normal concrete–as much as 1.0 mm/m–and due allowance must be made in all fixing details. Allowance must also be made for the fact that the tensile and impact strength properties of the material diminish with time. It is recommended that design stresses of 6.0 mN/m² for bending and 3.0 mN/m² for direct tension be used to ensure a safety margin of about five times the design condition when new. It is thought that this safety margin will fall to two or three times over the life of the component – which still leaves an adequate margin of safety. Developments are in progress to improve or overcome this loss of strength.

Fixing of GRC panels uses cleats or dowels in the same manner as that illustrated for precast concrete panels in Fig. 13.2.

13.4.3 Glass-Reinforced Polyester (GRP) panels

Liquid polyester resin is a thermosetting material (it hardens by a temperature-dependent process which cannot be reversed) and, when mixed with an activating chemical catalyst and glass-fibre reinforcement, cures to produce a durable compound material well suited to the production of cladding panels. It possesses characteristics of high tensile strength, low density, good corrosion and weather resistance and an ease of manufacture which makes short production runs economically viable. It is also possible to produce GRP panels in almost any colour, texture or shape desired. One main disadvantage is that, in common with many of the plastics, it possesses a low modulus of elasticity which means that it deflects more readily under load than other building materials.

To overcome this problem panels are designed with shaped profiles, ribbed construction or as a sandwich construction using a

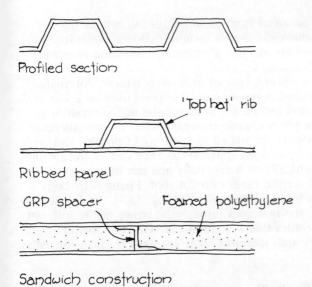

Profiled section

'Top hat' rib

Ribbed panel

GRP spacer   Foamed polyethylene

Sandwich construction

Fig. 13.3 Glass-reinforced polyester panels

foamed core (see Fig. 13.3). The foamed core can, of course, contribute to the thermal insulation of the panel as well as to its stiffness. A value of 0.5 W/m^{2}K is claimed for a GRP panel incorporating a 50 mm core of polyisocyanurate foam which is comfortably within current insulation standards.

The inherent ability of polyester resins to resist the effects of fire is inadequate for any buildings other than low structures (less than 15 m high) which are not near the site boundary. However, by the use of additives and fillers, the fire-resisting qualities of GRP panels can be improved to acceptable levels. Unfortunately, these additives and fillers reduce the weathering qualities of panels exposed to ultraviolet light. Various techniques are being developed to overcome this, e.g. coating the panel with an ultraviolet light-resisting treatment or by the use of intumescent polyesters (intumescent materials produce a carbonaceous foam when subjected to fire, which isolates the component from the flames).

The design of panel fixings is based on either a threaded plate moulded into the back of the panel into which a fixing bolt is screwed or a system of clamping the panel back to the structure, usually at the edges within the joint detail.

13.4.4 Metal cladding sheets and panels

Profiled metal sheet cladding has been in popular use as a means of enclosing a building for a long time; it was originally called corrugated iron and suffered many disadvantages. Its modern counterpart is far removed from these origins but retains the advantages of a lightweight simple form of building enclosure.

The metals used are steel or aluminium. Protection is provided to the steel by galvanising followed by a coloured and textured coat of p.v.c., acrylic or silicone enamel or fluoropolymers. Alternatively, a coating of a modified polyester can be applied over an epoxy resin base coat. Aluminium can be left in its natural milled finish or it can be anodised or given organic coatings similar to steel sheets.

The cladding sheets are ribbed to a variety of profiles and interlock to give a continuous appearance. They can be fixed with the ribs running vertically or horizontally and can be curved to run round corners or over the eaves of a flat roof. Fixing is by bolts through the sheets into cladding rails (see Fig. 13.4). In the case of sinusoidal profiles, the bolt goes through the crown of the roll, but with trapezoidal profiles this would have the effect of spreading and distorting the sheet and, therefore, they are fixed through the valley.

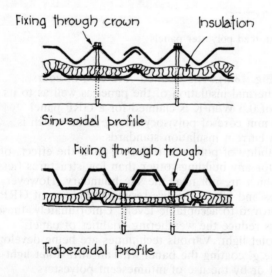

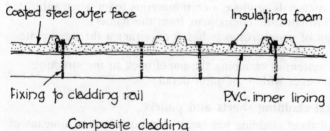

Fig. 13.4 Profiled metal cladding

Figure 13.4 shows a sandwich construction of outer sheet and inner lining sheet with an isulating quilt between the two to provide thermal insulation. The same objective can be achieved by a one-piece sandwich cladding with a protected steel outer lining, a polyisocyanurate foam core and a p.v.c. or similar inner lining.

Steel and aluminium protected and finished as described above can also be used to produce cladding panels in much the same way as the GRP panels described in section 13.4.3. Production calls for expensive equipment and, therefore, unlike GRP panels, a large number of standard panels must be produced to make the system economic. Sheet-metal panels can be used to create the complete wall enclosure as storey height units incorporating the windows or they can be used as undercill infill panels in a curtain walling system.

13.4.5 Curtain walling

A precise definition of this term is difficult because of the wide variety of systems in use, but generally any structural frame cladding which comprises a framework into which glass and obscured panels are fitted is considered to be curtain walling. Another term used in this context is 'patent glazing'. This arose in the nineteenth century and was prompted by the Crystal Palace. It refers to puttyless systems of glazing which were used in that structure and inspired an abundance of patents to be taken out on alternative designs. Modern methods use compressible sealing cords on to which the glass is pressed by clips screwed or sprung into place (see Fig. 13.5).

The traditional form of curtain walling, known sometimes as the 'stick system', is to attach metal-box section mullions to the structure–usually the edge of the floors-and between them fix transome rails to produce a framework grid which can be filled with panels as appropriate (see Fig. 13.5). Allowance must be made for thermal movement and for fixing tolerances, as shown in the illustration.

One of the problems to be solved by the designer of curtain walling is the penetration of rainwater at the joints. None of the water falling on the surface is absorbed, as in masonry construction, and tends to concentrate at the joints where, under the action of wind, it can penetrate the smallest opening. The reason it can penetrate the openings is because the external pressure in greater than the internal pressure. The technique now used to overcome this is a pressure-equalised wall construction. In this an air space between an outer rain screen and an inner air barrier is vented to the exterior so that the air pressure in the space is the same as that outside. This stops the water being forced through the joints. The joints in the outer skin are made with 'deterrent seals' and those in the inner skin are 'air seals' and must totally prevent any air leakage (see Fig. 13.6).

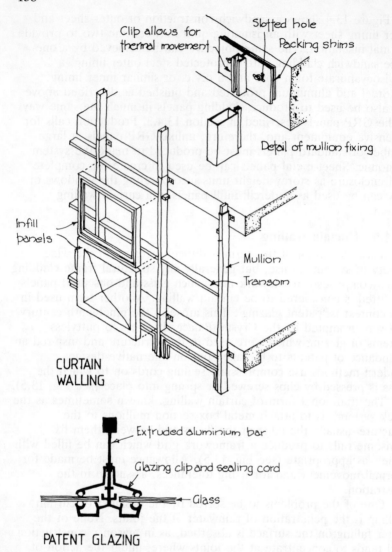

Clip allows for thermal movement

Slotted hole

Packing shims

Detail of mullion fixing

Infill panels

Mullion

Transom

CURTAIN WALLING

Extruded aluminium bar

Glazing clip and sealing cord

Glass

PATENT GLAZING

Fig. 13.5 Curtain walling patent glazing

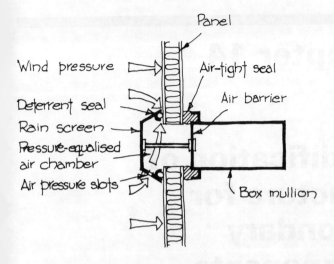

Labels: Panel; Wind pressure; Air-tight seal; Air barrier; Deterrent seal; Rain screen; Pressure-equalised air chamber; Air pressure slots; Box mullion

As the pressure in the air chamber is the same as the external pressure, rain is not forced past the deterrent seals

Fig. 13.6 The principle of pressure-equalised wall construction

Chapter 14

Modification of structure for secondary components

14.1 General provisions

With the complexity of modern building, one of the major tasks is the co-ordination of all the elements which go to make up the finished structure. The most detailed aspect of this work is the planning of the modifications which must be made in one element solely to allow another element to be satisfactorily incorporated in the building.

These provisions can vary from merely a few specially drilled holes to take, say, suspension rods supporting ventilation ducting, to a modification or strengthening of the whole structural system to fit round or to support a large piece of equipment.

In some cases, provisions for fixing secondary components can be made on-site in precisely the position required. Holes can be drilled in concrete and plugged to receive screw fixings, steel cleats and brackets can be welded to a steel frame to provide a support and bolt-fixing points etc. Where the position of these fixings is known in advance, it is much more efficient to form the holes or provide the cleats when the structure is being fabricated, but allowance must be made for a fixing tolerance to allow for on-site inaccuracies. In steelwork, these tolerances can be achieved by the use of slotted holes to give lateral adjustment and packing shims or washers to achieve a firm seating. In concrete work, a common solution is to form dovetailed recesses in the concrete which can receive purpose-made dovetailed anchors anywhere in their length.

On to the anchors are fitted washers to give adjustment off the concrete face.

14.2 Accommodating services

Pipes and cables can be run in vertical and horizontal ducts provided for that purpose throughout the building. The formation of these ducts and the need to consider the risk of fire spreading through them is dealt with in Chapter 6. The accommodation of ventilation ductwork is a more difficult problem than that of pipes and cables because of the sizes of the ducts and the need to make any changes of direction as gentle as possible.

The ideal shape for ventilation ducting is circular because, not only is it an inherently strong shape, but frictional resistance is less than in any other shape since a tube offers the smallest surface area for the volume enclosed. Large, circular ducts, however, do not always fit neatly into or up against the building structure and changes of cross-sectional area and junctions are difficult to form. For these reasons, rectangular section ducting is used, but the section must not depart from a square any further than absolutely necessary. The maximum ratio between the sides is usually taken as 1 : 3. With rectangular ducting there is a greater risk of the flat sides vibrating, due to the passage of air, and generating noise. Noise can also be due to the movement of the air itself.In both cases, the trouble can be minimised by restricting its velocity within the duct. This velocity can be between 5 and 10 m/s in main distribution ducts enclosed within a sound-insulating structure, dropping to about 2 m/s at the room inlets. With these values and a large building, the main ducts can be as much as 3 m^2 in cross-sectional area and require very careful thought in their design and accommodation. For instance, where ducts cross below the line of downstand beams is a situation calling for detailed analysis and consideration, because of the large total distance between floor slab soffit and underside of ducts. In this case, the heating engineer may be asked to re-route his duct or the structural engineer may be asked to avoid downstand beams at this point. It is now normal practice to provide suspended ceiling assemblies in commercial premises, above which are all the lighting and ventilating services, possibly also the heating, internal public address, TV surveillance and similar systems, plus fire-fighting sprinklers. All these must be accommodated in such a way that the depth of the ceiling void is kept to a minimum (to save on overall building height) and due provision or modification of the structure must be made to afford suspension points for the service runs and equipment, and the ceiling framework.

In the vertical direction, service ducts are generally provided and, where there are a lot of services to accommodate, the duct is of a sufficiently large size to affect the structure in two ways. Firstly, either the duct or the structural frame must be located so that they avoid each other, thereby assuring an uninterrupted run for the pipes, cables and trunking. Secondly, because the duct enclosure must be fire-resisting and, therefore, of substantial construction, it can be used as an anchor to stabilise the structure. This is explained in more detail in Chapter 11.

14.3 Accommodating equipment

Many buildings contain large items of heavy mechanical equipment attached to and supported by the structure and modifications must be made so that they can be positioned where required, fixed as necessary and the resultant loads safely carried. The most widely used equipment of this nature is a lift, but escalators, conveyors and cranes are all found fixed to building structures.

14.3.1 Lifts

Lifts vary from small document or goods lifts to large passenger lifts and those which can take a stretcher on a trolley or a car. In all cases, the need to provide a continuous vertical unobstructed shaft throughout the building and the loads imposed by the lift car, its occupants and the motor and winch will influence certain design decisions by the architect and the structural engineer. Very small lifts require nothing more than openings left in the floors and a fire-resistant enclosure fitted with fire-check doors. Larger, passenger-carrying lifts require a more sophisticated provision. In some building designs, the structure enclosing the lift is independent of, but tied to, the building structure. In others, the main structural frame is arranged to fit round the lift shaft, and the fire-resisting enclosure is constructed of beams at each floor level. As well as fitting the structure round the shaft, the normal building form must be modified at the top and the bottom of the shaft to form the lift motor room and the pit respectively. Lift motor rooms can present architects and structural engineers with difficulties because of their great height and because of the extension of the stanchions and the non-standard beam arrangements called for. The height arises from the need to allow for the lift car height, plus an allowance for over-travel, plus the lift motor room floor, plus the height of the lift motor room and the thickness of its roof. This can add up to as much as 8 m above the top floor served by the lift and means that the motor room roof protrudes well above the normal roof line (see Fig. 14.1). As can be seen from the illustration, the normal stanchions must be extended, others added at upper level and beams included at a level occupied by no other beams in the building. At the bottom of the shaft, the ground or basement floor structure must be interrupted and a lift pit constructed. This can be as much as 1.5 m below the lowest level served by the lift and allows for inaccuracies in the control of the stopping point of the lift car. Any structure built into the ground to this extent must be designed to withstand the pressure of the surrounding soil and to exclude ground water. Furthermore, the foundations may have to be placed at a lower level to avoid the pit floor and, consequently, the stanchions must be extended downwards to reach them.

Extending stanchions in the way described will probably mean increasing their size as well, because the slenderness ratio (the relationship between the least horizontal dimension and the effective length) will affect the load the stanchion can carry. An increase in size is also needed to allow for the additional loads involved to be carried with safety. Further modifications are needed at each landing doorway, as shown in the drawing, where the floor is required to project into the lift shaft by 120 mm and the beam supporting the floor must be splayed to avoid trapping any part of the lift car.

Instead of a motor and winding sheave, lifts can be operated by an hydraulic ram, set at one side of the lift shaft and connected to

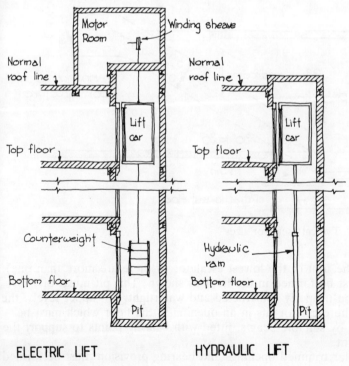

ELECTRIC LIFT   **HYDRAULIC LIFT**

Fig. 14.1 Lift details

the lift car by high-tensile steel chains. In this case, the load of the lift and its passengers is carried by the ram which bears on the floor of the lift shaft pit. The only strengthening of the structure required in addition to that of the pit floor, is to stiffen the lift shaft enclosure to take any horizontal forces imposed by the lift car guides. At the top of the lift shaft, the headroom required is much less than with an electric lift since no motor room is necessary (see Fig. 14.1).

14.3.2 Escalators

Escalators are installed in buildings where large numbers of people need to be moved through a limited height. The equipment is quite large and careful preparation must be made for its accommodation.

In most escalator designs, there is more of the machinery below the moving steps than above and the structure must be modified to fit this. Typical dimensions are shown in Fig. 14.2.

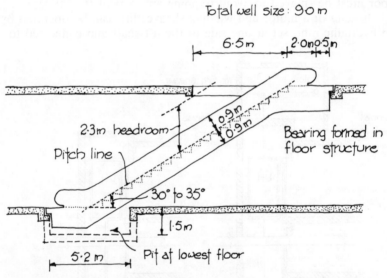

Fig. 14.2 Typical escalator sizes

At the foot of the lowest escalator (if there are more than one), a pit must be formed in the floor as shown. This pit must, of course, be properly constructed and watertight. On upper levels, the foot of the escalator sits in an opening in the floor which must be trimmed by suitable beams, fitted with bearing points to support the equipment.

Similar trimmed openings and bearing provision must be formed for the top end of the escalator and care must be taken to ensure that the opening is large enough to achieve the headroom clearance required. This size must be larger than would be required for a staircase because of the low pitch angle of the escalator. The beams trimming the opening must be designed to take not only the load of the adjoining floor, but the weight of the escalator machinery plus a superimposed load of as much as 4.0 kN/m^2 (depending on the building being served). The large size of the openings in the floors can present difficulties with the restriction of fire. For this reason, escalators are often enclosed in a fire-resisting shaft fitted with automatic fire-resisting doors, or are fitted with automatic sliding fire shutters which seal off the well opening at each floor level. This structure must be allowed for when designing the building layout and frame.

Another allowance to be made is that of access. Escalators are usually delivered to site in one piece and fairly late in the building programme. They are bulky and heavy and it is important that adequate means of transporting and handling them are provided so that they can be unloaded and placed in position.

14.3.3 Conveyors

Although, technically, the term conveyor embraces lifts and escalators, it is generally understood to refer to mechanical equipment installed to transport goods of various types from and to a number of fixed points or stations. With some types of conveyor, the stations can be on different levels. There are four main types of conveyor: belt, cable, chain and tube.

Belt conveyors consist of a continuous belt running round rollers or sprocket wheels. They are mainly for horizontal conveyance but, with suitable ribs across the belt, they can be used at an angle to serve an upper level. Very little modification of structure is called for with a belt conveyor. They usually stand on the floor but may be suspended from a floor soffit.

The loads carried are only light and, therefore, the only structural provision necessary is that required to afford fixing points. Building design is affected to a greater extent than construction because belt conveyors can only travel in straight lines and any change of direction involves transferring the goods from one conveyor to another. This imposes the need to plan all stations served by the conveyor in a straight line as far as possible.

A cable conveyor consists of an overhead wire stretched tightly across the building along which run travellers containing the goods. The traveller, which comprises a box or basket slung from a pulley and hanging beneath the wire, is either propelled by a spring-release mechanism or is wound along by a drag wire. The system is suitable for the conveyance of cash, letters and other documents, but only where these must travel regularly between a number of stations and one central main station. Firm anchorages are required for the ends of the wires to allow them to be tensioned sufficiently and, at intermediate support points, soffit fixings are required. If communication is between two rooms, suitable openings must be arranged in the structure of the partition but, apart from this, no other structural modification is needed.

Chain conveyors comprise an endless moving overhead chain to which is attached a series of hooks or baskets. It does not stop and goods are attached to the hooks or placed in the baskets as they pass. Discharge or collection can be manual or automatic. The chain can be inclined and can easily negotiate corners. Therefore, building design is not greatly affected, but a structural system which avoids deep downstanding beams, which would get in the way of the conveyor, is to be preferred. The floor structure should also be selected to provide a soffit which will afford a free choice of suspension-fixing positions, sufficiently adequate to support the large loads this conveyor is capable of handling.

Pneumatic tube conveyors have largely replaced the cable conveyors for the carrying of goods and documents between offices because they can travel in any direction, horizontally or vertically, and can be arranged as a ring main system, thereby allowing communication between any pair of stations. The system uses a steel or p.v.c. tube, between 50 and 150 mm diameter, or up to about 400×150 mm in rectangular form. Inside the tube are placed carriers which fit closely to the tube shape. Around them are strips of felt to give an airtight seal between the carrier and the tube. Air is pumped out of the tube ahead of the carrier, causing it to be propelled along by the atmospheric pressure behind it. A selector switch, activated by the passing carrier, operates a diverter flap to deliver the goods to the appropriate station.

One of the attractive features of this system is that it can be neatly concealed but this facility, plus the need to keep bends to as large a radius as possible, means that careful planning and modification of the structure will be required. The planning must ensure that the tube can be run to each station and that suitable horizontal and vertical ducts are provided as well as a plant room to accommodate the vacuum pumps and controls. Structural considerations will be concerned with the layout of beams and columns, in relation to the tube runs, the provision of suitable openings through which to thread the tube, the provision of fixing points and any special requirements of the plant room in the way of load to be carried or vibrations to be suppressed.

14.3.4 Cranes

In many industrial buildings and warehouses, goods are moved by an overhead travelling gantry crane. This is a specialised type of conveyor which can handle large heavy objects and move them to any position within a specific floor area. The equipment consists of a lifting hook suspended from a hoist which runs on a beam or gantry spanning the width of the building.

The gantry is attached to trolleys mounted on rails which run the length of the building and are fixed to the structure (see Fig. 14.3). The requirements of this equipment impose considerable influences on both the building design and its structure. The design features are that the plan shape of the building is preferably a long narrow rectangle (thereby achieving the desired floor area but keeping the span of the gantry within economic limits) and the sectional shape must provide adequate clearances each end and above the gantry. The maximum span usually encountered is 18 m and clearances can be as much as 0.36 m end clearance and 3 m height clearance. There is no practical limit to the length of the building but a very long travel for the gantry may not be efficient in operating time.

The loads placed on the supporting members of the building and their foundations can be large and, since it is a moving mass, complex. Travelling gantry cranes can have a lifting capacity of as

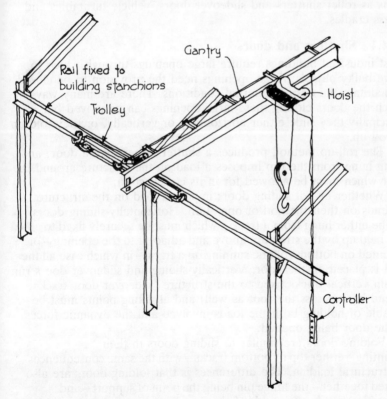

Fig. 14.3 Typical gantry crane

much as 50 tonnes and can be required to be mounted high above the floor to allow transporting clearance over stored goods. Obviously there is a direct downward load on the rails, due to the mass of the gantry and the trolley plus the lifted goods. This is concentrated at a point on each rail, but it must be taken as any point in the entire length of the rail. In addition, the dynamic nature of the crane load transmits horizontal thrusts both along and at right angles to the rails. The combination of these forces acting high up on the structural stanchions can present the design engineer with some difficult problems, resulting in the final structure and foundations being considerably different to that which would have obtained had there only been the building loads to carry.

14.4 Accommodating secondary components

Many industrial and commercial buildings incorporate large, mechanical secondary components which must be allowed for in the design of the building and its structure. Typical of these are such items as roller shutters and slideover doors, vehicle turntables and access cradles.

14.4.1 Shutters and doors

Most industrial buildings require large openings through which to move bulky objects; these openings need the provision to be closed to maintain the desired internal conditions. There are many ways in which the doors closing these large openings can be moved but principally they slide either horizontally or vertically, or they fold up or roll up.

The roll-up method produces a shutter rather than a door, and being hung from the top imposes a load on the structure around the door which must be allowed for in its construction.

Whether or not sliding doors impose a load on the structure depends on their method of operation. Horizontally-sliding doors can be either hung from a track – which must be securely fixed to and held up by the structure above and adjacent to the opening – or mounted on bottom rollers, running on a track – in which case all the load is placed on the floor. Vertically-sliding and slideover doors run within vertical tracks fixed to the structure (slideover door tracks extend back across the roof as well) and all fixing points must be capable of holding both the loads involved and the dynamic forces of the door being opened.

Folding doors are similar to sliding doors in their mounting – either top or bottom tracks – with the same consequences of structural loading. The differences is that folding doors are also hinged together – the hinge pin being the point of support – and concertina within the width of the opening.

14.4.2 Vehicle turntables

These are found in buildings wherever a lot of vehicular manoeuvring is required in a confined space, and consist of a large circular steel deck structure mounted on a centre bearing and outer support rollers, set flush with the floor. The vehicle is driven on to the turntable which is then rotated, either manually or by an electric motor, until the vehicle is facing the desired different direction, when it is driven off.

Capacities range from light-duty, between 3.35 and 4.57 m diameter or 3 to 10 tonnes, through medium-range to heavy-duty turntables which are 6 m and over in diameter and can be designed to carry any capacity over 10 tonnes.

The floor must be accurately constructed with a circular hole to fit the turntable precisely. Below the hole there must be a pit between 267 and 550 mm deep, depending on the size of the installation. The base of this pit has to carry the dead and dynamic loads of the turntable and the vehicle and must, therefore, receive careful attention in its design and construction.

14.4.3 Access cradles

For maintenance purposes, many large buildings are equipped with a set of steel tracks mounted on the flat roof on which runs a trolley supporting arms which extend out beyond the building face and provide suspension points for cradles in which maintenance workmen can be carried.

The tracks are mounted on roof anchor points which are often concrete blocks, cast on the roof with bolts set in the top of them to receive the track fixings. The roof weathering finish is turned up the blocks and sealed all round them to prevent the rain getting in. These anchorage points must be positioned where the roof has the necessary strength to support them or, conversely, the design of the roof must incorporate the necessary strengthening.

Chapter 15

Commercial and industrial service requirements

15.1 Artificial lighting

The demand for artificial lighting is probably as old as the controlled use of fire by early man and, indeed, the means by which people supplemented the daylight or substituted for natural lighting at night has been by means of small portable or non-portable 'fires' in the form of rush lights, candles, oil lamps and gas lamps. With the advent of the incandescent filament lamp and, more recently, the fluorescent tube and various types of discharge lamp, we now have at our disposal sources of light which are much more powerful and accurately controllable. Because they will constantly produce exactly the conditions we require, they now substitute for natural light during the day time as well as at night.

The initial requirement of any lighting arrangement is that it shall produce sufficient illumination for the task in hand, and many publications contain tables giving recommended levels of lighting for different situations. In this chapter it is explained how to ensure these recommendations are met. Now that we can, with ease, achieve intensities of light adequate for even the finest work, the emphasis is moving increasingly towards the quality of that light. The colour and variety of the lighting, the avoidance of glare and the way the light brings out the form and texture of objects are subjects receiving much more attention today than they have done previously.

15.1.1 The design balance of windows and artificial lights

In the past, the design policy was to make windows as large as possible to ensure an adequate supply of natural light by which the occupants of the building could work. In some buildings the intensity of the light entering when the sun shone was so excessive that solar-control film had to be applied to the glass or the glass replaced altogether with tinted glazing. Alternatively, the windows were (and still are in many of these buildings) fitted with shutters, curtains or blinds. Whatever arrangement was used, the intention was to restrict the over-provision of glazing. At the same time, the heat of the sun created intolerable room temperatures. In the winter, the shielding against the light was not required – indeed, on a dull day, artificial lighting could be needed over the work places remote from the windows – but the room temperature would drop and strong, cold down draughts would be set up, again producing uncomfortable working conditions and high heating bills.

Designers have now realised that large windows are not usually economically advantageous. Much of the time, artificial lighting is needed even with the large windows and there is a big heat loss which, in many cases, wastes more energy than is saved by not using artificial light.

The conclusion to be drawn from this is that it would be best to omit windows altogether and, from the point of view of simple economics, that is probably true. But buildings are for people. Everyone likes to see out of the space in which they are enclosed and, for those unfortunate enough to suffer from claustrophobia, this liking becomes a psychological necessity.

The present tendency is towards much smaller windows, and a complete reliance on high-efficiency artificial lighting fittings, now available, as the source of illumination. This achieves a better energy balance, and in more sophisticated systems the heat of the lights can be recovered to be used for space heating around the room perimeter.

15.1.2 Lighting criteria

The design criteria on which an artificial lighting scheme would be based are concerned with either the quantity or the quality of the light.

Whether the building occupant is relaxing or working, the critical lighting factor which affects him is the quantity of light where he is or where he wants to see, not the quantity at source. For this reason, lighting systems are designed to produce a specified illumination at the working plane.

This plane may be a desk or bench top, it may be the page of a book or it may be a sloping or vertical surface such as a drawing board or display board. The values are expressed in lux and range from 30 – 50 in a hospital ward to 100 in an auditorium, 300 generally in a wood machine shop, 500 for general clerical work, 750 on a drawing board and 1000 on a fine assembly workshop bench (see Ch. 17).

Light quality is a complex matter concerning the distribution number, brightness and design of the light fittings. These should be combined into a balanced arrangement to produce the desired illuminance without strong contrasts giving rise to glare. At the same time, there should be sufficient shadow to define the three-dimensional form of objects.

15.1.3 Lighting fittings

The term 'luminaire' has now superseded 'lighting fitting' in the lighting engineer's vocabulary. Luminaires can be classified in several ways, depending on the interests of the classifier. Reference can be made to the type of light source – tungsten filament lamp (technically an incandescent lamp and commonly the normal 'bulb'), fluorescent lamp, metal vapour lamp (orange street-lamp type – in this case sodium vapour), discharge lamp (in which the light is produced by the excitation of a gas) etc. Classification can also be made on the basis of use – decorative fittings (appearance is important as lighting ability), general utility (the main bulk of luminaires) and special fittings (spot lights and similar highly directional units). A more technical classification is concerned with the direction of light emitted by the luminaire and can range from indirect, where all the light is projected upwards, to direct, where all the light is projected downwards (see Fig. 15.1).Many manufacturers define the light distribution from their luminaires by a reference of BZ followed by a number between 1 and 10.

The BZ is an abbreviation of British Zonal Classification and the number refers to the distribution of downward light. A BZ1 luminaire throws most of its light down through an angle of 45° each side of vertical, whereas most of the light from a BZ10 luminaire is given out between 45° and horizontal (see Fig. 15.1).

Most luminaires are designed to shield the lamp from direct view to reduce the possibility of glare – Figure 15.2 illustrates typical fittings. The result of this shielding is a certain loss of useful light, some of which may be diverted upwards, some downwards and some is lost. If the light from the luminaire is divided by the light from the lamp, the result is the light output ratio (see Fig. 15.3).

With the use of suspended ceilings has come the luminaire designed to fit into the ceiling grid or to take advantage of the deep ceiling void. Modular fittings (see Fig. 15.2) are made to the same dimensions as ceiling tiles, usually 600 × 600 mm, and consist of an inverted metal box containing the lamp or lamps which fit into the ceiling suspension system, and a diffuser clipped into it of either a

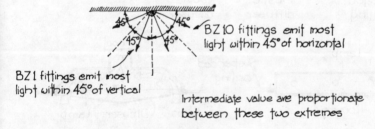

Type of luminaire	Percentage of upward light
Direct	0 to 10 %
Semi-direct	10 to 40 %
General diffusing	40 to 60 %
Semi-indirect	60 to 90 %
Indirect	90 to 100 %

CLASSIFICATION BY UPWARD LIGHT PERCENTAGE

BZ 10 fittings emit most light within 45° of horizontal

BZ 1 fittings emit most light within 45° of vertical

Intermediate value are proportionate between these two extremes

BRITISH ZONAL (BZ) CLASSIFICATION

Fig. 15.1 Classification of luminaires

translucent material or a grid of louvres. 'Down lighter' fittings are smaller than the ceiling tiles and consist of a circular bezel fitted round a hole cut in the ceiling tile behind which is a can lined inside with black fins and supporting an internally-silvered spot or flood lamp. This luminaire extends some distance up into the roof void and produces a strong, but more or less invisible, localised light source.

Extractor fittings are designed to combine the function of a luminaire and an air-conditioning extract point (see Fig. 15.2). This

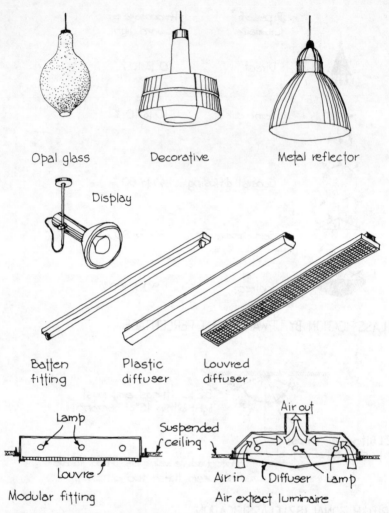

Opal glass   Decorative   Metal reflector

Display

Batten fitting   Plastic diffuser   Louvred diffuser

Lamp   Suspended ceiling   Air out

Modular fitting   Air in Diffuser Lamp

Louvre   Air extract luminaire

Fig. 15.2 Typical luminaires

arrangement has the advantage of removing the convection heat of the lamps, thereby controlling the room temperature and cooling the lamps; this extends their life and provides a supply of warmed air which can be used as an energy source for space heating. The associated ducting makes a suspended ceiling necessary, but this type of fitting is generally used in conjunction with an air-conditioning system which also needs a suspended ceiling.

Some suspended ceilings are not a continuous closed surface but consist of louvres or a grid of metal leaves or may be a translucent film. In all these versions, light can be transmitted through the

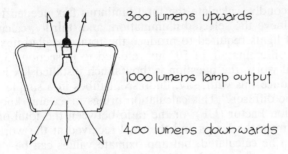

300 lumens upwards

1000 lumens lamp output

400 lumens downwards

Light output ratio:

$$LOR = \frac{\text{Light from luminaire}}{\text{Light from lamp}}$$

$$= \frac{300 + 400}{1000}$$

$$= 0.7 \text{ or } 70\%$$

Upward light output ratio

$$ULOR = \frac{300}{1000} = 0.3 \text{ or } 30\%$$

Downward light output ratio

$$DLOR = \frac{400}{1000} = 0.4 \text{ or } 40\%$$

Fig. 15.3 Light output ratio

ceiling – indeed, this may be the reason the particular ceiling design was chosen – and, therefore, very basic luminaires are needed. With many installations, these consist of a simple batten fitting and bare fluorescent tube mounted above the ceiling, at a height of not less than two-thirds of the spacing of the tubes.

15.1.4 Lighting design

To ensure that the recommended level of illumination is achieved on the working plane, it is necessary to calculate the number and rating of the lamps to be installed. This is a task usually undertaken by a lighting engineer, but the following method will serve for preliminary purposes.

The three steps are: firstly, select the illumination required, i.e. the luminous flux per m² of work surface, and the type of luminaire to be used; secondly, calculate the total luminous flux needed to be installed to achieve the selected illumination; and, thirdly, calculate the number of lights required to produce the total installed flux.

Taking the illumination of 500 lux referred to in section 15.2 for general clerical work, the question is 'how much installed lux is needed to produce this with, say, fluorescent tubes in a simple enclosed plastic diffuser?' This calculation brings in a value known as the Utilisation Factor (UF), or the ratio between the total of the luminous flux from the lamps and the total received at the working plane. This can be calculated, but approximate values can be obtained from manufacturers' published tables; in the example being studied, the UF could be taken as 0.49 and can be expressed by the fraction:

$$0.49 = \frac{\text{Flux received on the working plane}}{\text{Total installed flux}}$$

or

$$\text{Total installed flux} = \frac{\text{Flux received on the working plane}}{0.49}$$

$$= \frac{\text{Illuminance} \times \text{Area of working plane}}{0.49}$$

Assuming the room to be 10 m long and 5 m wide, this becomes:

$$\frac{500 \times 10 \times 5}{0.49}$$

$$= 51\,000 \text{ lm}$$

Reference must again be made to the manufacturer's catalogue to find the lighting design lumen output (LDL) of the lamp. Assuming a 65 W fluorescent tube, the LDL value would be about 4400 lm. This means that the luminaire will have an output of this value, after allowing for the fall which always occurs with time. Dividing this LDL into the required total installed lux gives the number of luminaires needed:

$$\text{Number of lamps} = \frac{51\,000}{4400}$$

$$= 12$$

It then remains to arrange these 12 luminaires uniformly over the area of the room to produce an illumination which is as even as possible. In this connection, a variation down to 70 per cent below the maximum is considered acceptable. Figure 15.4 shows a possible arrangement of three rows of four luminaires. With this, the spacing is 1.25 m between the rows and 2 m between the luminaires. The

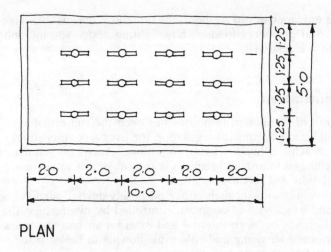

PLAN

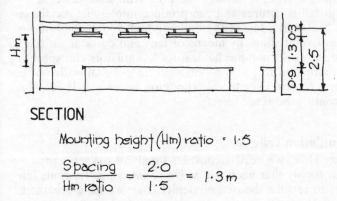

SECTION

Mounting height (Hm) ratio = 1·5

$$\frac{\text{Spacing}}{\text{Hm ratio}} = \frac{2·0}{1·5} = 1·3\,\text{m}$$

Fig. 15.4 Lighting design layout

height needed between the working plane and the luminaires to achieve 70 per cent distribution (the mounting height, Hm) can be found from manufacturers' data, which usually quotes a spacing to mounting height ratio for each pattern of luminaire.

For the plastic diffuser type of fluorescent tube luminaire being considered, this ratio is of the order of 1.5 : 1, i.e. the spacing to be not more than 1.5 times the mounting height. With a maximum spacing of 2 m, this gives a mounting height of 1.3 m (see Fig. 15.4). Taking the working plane as 900 m above floor level and allowing 300 mm between the centre of the luminaire and the ceiling, gives a total required floor to ceiling height of 2.5 m.

There are a number of other factors such as the anticipated maintenance and the reflective qualities of the room surfaces which the lighting engineer would include, but the above calculations give an initial layout and serve to show how lighting design spacing and room height are related.

15.2 Ventilation

The provision of ventilation in houses has been the subject of national and local government legislation for over a century in an endeavour to achieve healthy living conditions. Adequate ventilation in commercial and industrial premises is the subject of rather more recent legislation and recommendation.

Natural ventilation, through either purposely-devised air inlets or cracks around windows and doors, in controlled by two factors: the pressure difference between internal and external air and the 'stack effect' of warmed air rising and colder air flowing in below it to take its place (in the same way that air rises up a chimney stack thereby creating a draught through the fire). The stack effect is more acute in tall structures and can produce intolerable conditions of excessive heat at the top of the building and cold at the bottom.

Mechanical ventilation, by means of fans and ducts or, in single-storey industrial work, non-mechanical roof ventilators can substitute for natural ventilation for, say, rooms with no external walls or ones in which a fine control over the air condition is needed. It can also be used to control the stack effect.

15.2.1 Ventilation criteria

For us to stay alive, we need a constant supply of oxygen and, apparently to supply that need, we strive to introduce fresh air into our buildings to replace the oxygen-depleted air we have breathed. Essential though it is, the oxygen content of the air does not form the basis of any standards of ventilation. This is because long before the oxygen reaches a dangerously low level, other air conditions cause acute discomfort and even death.

Some of the criteria are concerned with the quality of the air and what it contains; others are involved with the maintenance of physical comfort and health. Of these latter criteria, the most important are humidity and air movement. These two are closely linked and together control the rate of evaporation of moisture from our bodies. Without any evaporation, we would rapidly become over-heated and expire; with too much evaporation, our bodies become dehydrated.

Humidity is caused by water vapour present in the air and can be expressed in terms of grams of water per cubic metre of air. There is a finite amount of water vapour which can be contained by the air; the limit is called saturation point, and the nearer the air is to being saturated the less the evaporation from our bodies and the greater our discomfort. The amount of water vapour which the air can contain depends on its temperature. Since this is variable the ventilation criteria is based on relative humidity (RH). Relative humidity is expressed as a percentage and represents the quantity of water vapour present compared to the maximum possible at that temperature. For comfort, we need an RH of between 40 and 60 per cent . If the temperature falls, the amount of water vapour the air can hold also falls. This often happens at night. At the same time, the amount of vapour present remains the same, thus causing the relative humidity to rise. Eventually the RH reaches 100 per cent, condensation occurs and dew forms. If this situation arises internally as at a window pane, the result is referred to as condensation and can result in misted windows or damp walls.

Air movement occurs quite naturally with the movement of the building's occupants and equipment as well as by leaks around doors and windows. In extreme situations, the natural movement is enhanced by the use of fans. Movement of air across the skin is required to produce a feeling of comfort; however, too much in the wrong situation or air at a low temperature can produce discomfort. The effect is to carry away the air against the skin which has become moistened with perspiration, replacing it with drier air thus allowing the evaporation to continue. It is this process which makes the air stream from a fan on a hot summer day seem to have a cooling effect, even though the air temperature is no different to that in the rest of the room.

Body odours, fumes, smells, dust, products of combustion and bacteria are found suspended in all the air around, and can be seen when drifting through a beam of sunlight. Our respiratory system is designed to deal with a lot of these fine particles, provided that they are at a fairly low concentration. Without ventilation, many of these would steadily increase, body odours would become stronger – especially if the temperature also rose – dust would become denser – much 'dust' is actually the human skin cells we are constantly shedding – and bacteria would proliferate. The purpose of ventilation is to carry away this smelly, dusty, stale air and replace it with fresh external air. In some cases, such as where fumes or smoke are caused, ventilation should take place at source with hoods or extractors positioned to catch them as they are produced.

15.2.2 Ventilation methods

The oldest form of ventilation is a hole in the wall, which also served to admit light. As a means of controlling internal conditions it was useless, but did ensure that every occupant received an adequate supply of fresh air!

The method most commonly used now has only progressed beyond this medieval crudity by introducing a hinged flap to close

160

the opening or to allow a restricted flow of air to enter. These hinged flaps are the windows we open and close to ventilate our rooms. They do ensure that the air is moved and that it is changed, but whether fresh air is being allowed in through the window or stale air being extracted through it – to be replaced by stale air from other rooms – is a matter of relative internal and external air pressures. These are as variable as the wind and its direction.

As with lighting, we now have technologies which can be substituted for natural phenomenon with a much greater degree of control and can ensure that air passes in through the inlets and out through the outlets and, furthermore, can be caused to do so in rooms which are remote from any external walls. These technologies consist of a variety of fans and impellers, which are mechanically driven and are linked to ducts through which the air passes, or a number of non-mechanical roof ventilators for use in industrial types of buildings (see Fig. 15.5).

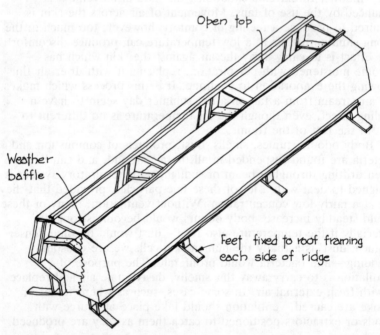

Fig. 15.5 Typical roof ventilator

As well as allowing rooms to be sited away from the external walls, mechanical ventilation can also ensure that comfort conditions are maintained in rooms containing a lot of people in a small volume; closely control the environment where critical standards of humidity or dust-free air are demanded; allow adequate ventilation in situations where natural ventilation through windows is not possible.

15.2.3 Affect of ventilation on design

Natural ventilation through openable windows and doors or air bricks requires, firstly, that the room so ventilated is against an external wall; secondly, that the ventilation opening is of adequate size; and, thirdly, that it connects to an open space sufficiently large to provide enough fresh air.

The location of the room within the plan is a matter of careful and, at times, wasteful arrangement. The wastefulness can arise when a corridor is introduced solely to reach, say, a WC sited between larger rooms (see Fig. 15.6). Adequate ventilation opening is generally taken to be one-twentieth of the floor area and the open space outside the window is required, by the current Building Regulations, to be generally 3.6 m deep and 3.6 m wide at the outer edge, narrowing towards the building. This open space must extend, uninterrupted, to the sky.

Mechanical ventilation relieves the designer of the constraints of locating rooms against the external wall, particularly small service

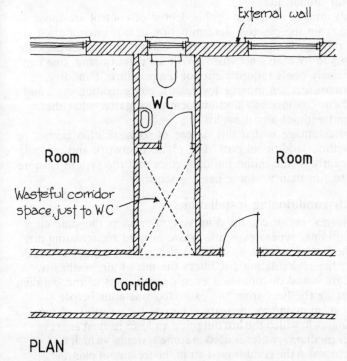

Fig. 15.6 Wasteful plan to achieve ventilation

rooms like bathrooms and WCs. Furthermore, the ventilation air can be drawn in and the stale air expelled at any point in the enclosing fabric, wherever is suitable. This makes the siting of the building easier since the air volume and condition immediately outside the room to be ventilated are no longer critical.

15.3 Air conditioning

The first step in providing contemporary facilities within a building is to install space heating. It is then discovered that this causes less than desirable air and comfort conditions and condensation on or within the external fabric and so controlled ventilation is introduced. The next logical step is to combine the two and to use the ventilating air to provide space heating as well. This, with the addition of filtration and humidity control, is air conditioning.

Many combined heating and ventilating systems claim to be air conditioning, but to qualify for this name it is necessary to do as the title describes, condition the air. The system must produce an environment in which not only is the temperature and quantity of the air correct for us but also the quality of the air is the best condition for human life.

The advantages in achieving this degree of control are those which arise from the positive determination of our environment instead of relying on the widely varying natural conditions plus the additions made by man's activities. With air conditioning, one can specify precisely one's requirements of temperature, humidity, oxygen content, carbon dioxide level, dust concentration etc., and maintain them continuously and uniformly no matter what the weather, and without any draughts.

The disadvantage is that this degree of sophistication cannot be achieved without additional cost. The plant, ductwork and controls add significantly to the initial building costs and the system is more expensive to run than a simple heating scheme.

15.3.1 Air-conditioning installations

In simple terms, an air-conditioning system extracts the stale air from the building, separates it into waste air and recirculating air, recovers the heat from the waste air which is then exhausted, adds fresh air to the recirculating air, filters the mixed air, washes it, dries it, heats it and distributes it evenly to all parts of the building.

Recovering the heat from the extracted waste air before releasing it is an economy measure achieved by a variety of types of heat exchangers in which the hot outgoing air loses its heat energy to a transfer medium – water is used in some systems – which is then taken to warm the conditioned air in the treatment plant.

Air-conditioning plant is usually made up on a modular basis,

that is, each stage of treatment is performed in a standard-sized
component and the required number and types of components are
bolted end to end to form a long, in-line treatment unit (see
Fig. 15.7). The usual sequence of modules is: a mixing box, a filter
unit, a pre-heater, a cooling coil, a spray washer, a re-heater and a
circulating fan.

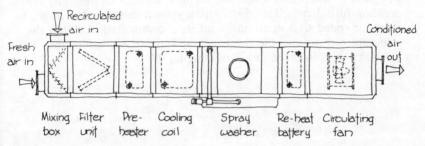

Fig. 15.7 Air-conditioning plant

The mixing box receives both the fresh and the recirculating air
and contains dampers to control the quantity of each. The filters
used in the filter unit can be a simple porous paper or fabric screen
filter but, since these increasingly restrict the air flow as they
become clogged with dust, alternatives have been developed. An
automatic roll filter is one such alternative and comprises a slowly
moving endless mesh belt passing through the air stream and into an
oil bath. The oil removes any dust caught and coats the mesh with
oil to catch more dust on its return. The most efficient filter is the
electrostatic type in which the dust particles are drawn to fine wires
by electrostatic attraction created by a high-voltage current passed
through the wires. In all cases, regular maintenance is necessary to
remove, clean or renew the dust-catching medium.

The pre-heater and cooling coil modules are used to adjust the
air temperature to ensure that the desired humidity levels are
achieved when the air passes through the spray washer. The re-
heater raises the temperature to the required level for space heating
and the circulating fan impells the air round the system. With some
in-line systems, an axial flow fan is used (one in which the fan
rotates in the air stream) but, in many cases, a centrifugal fan may
be employed in which the air enters the centre of the fan, is
impelled outwards by centrifugal force and delivered at right angles
to the direction of intake. The equipment is large, but not heavy,
requires an adequate supply of fresh air and the means to discharge
stale air.

For these reasons, the most convenient location is on a flat roof
and so air-conditioning modules are manufactured as double-skinned
components, using galvanised or stainless steel, with a rigid thermal

insulating foam core between the skins.

Distribution of the conditioned air and collection of the stale air is effected through circular or rectangular ducts. These must be sized and constructed in such a way that any noise generated by the air-conditioning plant is not conveyed into the rooms, nor is any generated by the ductwork itself. The outlet grills, too, must be designed to assist and direct the air flow without causing any noise-creating turbulence. This distribution system can be quite large, to keep air speed to a minimum, and its accommodation within the structure can present the designer and builder with problems. These are dealt with in more detail in Chapter 14.

Chapter 16

Fire control in commercial and industrial buildings

16.1 The cause and effect of fire

Ever since primitive man brought his fire inside his hut, his descendants have been trying to overcome the consequences of letting it get out of control. To understand how to contain a fire, either a deliberate heating fire or an accidental conflagration, it is necessary to consider how a fire burns and spreads through a building, and the hazards which develop.

A fire requires three contributory factors to start and to continue: heat, combustible material and a supply of oxygen. Remove any one of these and the fire goes out. Thus, spraying with water cools the burning material and reduces its combustibility, smothering the flames with a blanket of foam excludes the oxygen; in either case, the fire is extinguished. All fires start as small primary fires – an overheated electric cable, carelessly thrown cigarette end or match. If this continues unchecked, the heat of the fire raises the temperature of nearby material, causing it to give off combustible gases which then commence burning. In time – about 15 minutes – material over a wide area or complete room is generating combustible gas which suddenly ignites into an intense fire. This is known as 'flash-over'.

Once flash-over has occurred, the fire has reached serious proportions and the building structure will start to be affected; non-combustible materials, like steel, will become weakened and collapse; concrete members may disintegrate due to temperature differences and lose strength and the building becomes structurally unstable. While this is happening, what of the occupants? The first threat to life is the heat if the occupant is near to the fire. Principally, the breathing of hot air which sears the lungs. There is also a severe depletion of oxygen, leading to asphyxiation.

Occupants remote from the seat of the fire will be threatened by smoke which, if sufficiently dense, will make breathing difficult and finding a means of escape impossible. When a building is burning fiercely, it is not only hazardous to the occupants and contents, but also to adjacent properties and their occupants. Containment of a fire is the subject of much technological development ever since 1666 when a fire in a baker's shop in Pudding Lane eventually spread to destroy five-eighths of the City of London.

Spread of fire is by conduction, convection and radiation, plus the spread of flame across the surface of combustible materials. The heat propagation phenomenon which affects adjoining property to the greatest extent is radiation: heat escaping from doors and windows which can raise the temperature of neighbouring structures to a point where the least spark will cause another fire to start.

Building design in relation to fire must, therefore, take into account the siting of the type of structure in relation to other buildings, the internal planning to provide means of escape and fire control, the containment of the fire within the building by compartmentation and fire and smoke stops, the protection of the building structure and the fire resistance of the fabric generally.

16.2 Siting and design of buildings

The principles behind the Building Regulations in connection with controlling the risk of fire spreading are explained in detail in Chapter 9. They consist of defining a distance to be observed between the building and the site boundary in relation to the extent of 'unprotected areas' (i.e. windows), which present the hazard of radiated heat affecting a neighbouring property.

The Regulations are also designed to ensure that the building to which control is being applied is not at risk from a neighbouring fire due to there being combustible external facings near to where a neighbouring fire could start.

The design team must also consider the location of high fire risk buildings such as boiler houses, oil storage areas, paint stores and any storage building containing combustible materials. These should be segregated to ensure that other buildings are not seriously endangered.

A further consideration is access for fire-fighting equipment, both around the site and up to the seat of the fire if tall buildings are involved. Fire-fighting appliances can be quite large and heavy when laden with men and water, and properly constructed, clearly defined access routes from the public road must be provided. Any gate must be at least 3 m wide, the access route must be not less than 3.6 m in width with 3.6 m minimum clearance height. Changes in direction and junctions must accommodate a vehicle turning circle of 21 m diameter and the carriageway construction must be capable of supporting 18 tonnes. Adjacent to the buildings, the road, or a hardstanding, must be at least 4.8 m wide to allow the jacks of turntable ladder appliances to be set down.

The fire appliance access routes may be the normal service roads provided to serve the building complex or, if such roads are not suitable or do not reach all buildings, additional provision must be made with, for example, specially designed precast concrete road paving units which provide a running surface for occasional use but also allow grass to grow through holes provided for the purpose, thus making the road harmonise with the surrounding grassed areas.

16.3 Internal planning for escape

The basic principle in the design of any building is that it must be possible for any occupant to escape from fire or smoke by his own unaided efforts. To do this, the building plan must be so arranged that any person confronted with a fire can turn away from it and, proceeding in an opposite direction, make his way to the safety of either a protected staircase within the building or outside the building. This implies that alternative means of escape must be placed at the extremities of the building. Legislation controlling the design of commercial and industrial buildings in relation to fire is contained in several Acts of Parliament, particularly the Health and Safety at Work Act, the Building Regulations, the Factories Act, the Offices, Shops and Railway Premises Act and the Fire Precautions Act. Reference is made in them to Code of Practice 3: Chapter IV: Part 2: 1968 in relation to shops and departmental stores and Part 3: 1968 in relation to office buildings. In the Inner London boroughs, the Code of Practice, Means of Escape in case of Fire, produced by the Greater London Council, applies to all buildings except flats and maisonettes (where the recommendations of CP3: Chapter IV: Part 1: 1971 are adopted) nor to a group of building types such as nursing homes, institutional buildings, shopping malls and inflated structures (which are deemed to require individual scrutiny).

The principle that escape must be achieved by an occupant's unaided efforts has been established because it is not possible to rely on rescue by the fire brigade in all cases, or in time. There is an inevitable delay before the fire brigade arrives, parked vehicles may obstruct the positioning of ladders, fully air-conditioned buildings with non-openable windows add further problems and any upper parts of the building set back from the lower storeys or

168

any floors over 30 m above the ground (about the 11th storey) are inaccessible by any external means. Even at 12.6 m the wheeled escape ladder, normally carried, is at its maximum extension. To provide for this, means of escape design involves the siting and design of protected vertical shafts throughout the height of the building, containing staircases, and the direct and travel distances to them.

16.3.1 Protected shafts

The enclosure surrounding the fire escape stairways must possess a fire resistance related to the building use and size: this can vary from half an hour to four hours. Any doors in the enclosure must open in the direction of escape and be half-hour resisting if the building is an office or a shop, and half the required period of resistance of the enclosure if the building is a factory, place of assembly or storage building.

It is also necessary to prevent these protected shafts from filling with smoke (people would not then use the stairs even though the enclosing shaft is fire resistant). Two techniques exist to deal with this. Firstly, to remove any smoke leaking in by good ventilation through either opening windows or a roof vent or, secondly, to pressurise the shaft to prevent the smoke from entering at all. The first technique is the more commonly adopted and in its simplest form relies on normal manually-operated windows at each storey or landing level. More sophisticated installations have windows which are automatically opened when the fire alarm is sounded. Where it is impracticable to have openable windows, a permanent ventilator can be fitted at the top of the shaft which CP3 recommends should have an area of at least 5 per cent of the area of the enclosure. As the ventilation will be affected by wind direction and pressure, its efficiency is questionable and, therefore, the pressurising technique has been developed.

Pressurising the escape stairway not only prevents it from filling with smoke, but also increases the fire resistance of the doors by ensuring that any air leaking round the door is cool air from the stair well and not hot gases from the fire. In some installations the pressurising is constant, in others it starts when the fire alarm is activated. The necessary pressure is almost unnoticeable (a maximum of 13 N/m^2) and is readily maintained by a small fan.

16.3.2 Direct and travel distances

Direct distances are measured from the furthest point of the building to the protected shaft, within the perimeter of the building, ignoring internal partitions but not passing through the stairway enclosure. Travel distances are the actual distance a person would need to travel in negotiating doorways, corridors and fixed obstruction (see Fig. 16.1). In a building where people work, the

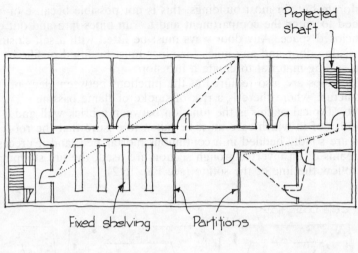

Fixed shelving   Partitions

Direct distance   ....
Travel distance   ---------

Fig. 16.1 Direct and travel distances

GLC Code of Practice states that the direct distance should not exceed 30 m – unless it is a high fire risk area, in which case the distance should not exceed 12 m – and the travel distance should not exceed one and a half times the permitted direct distance.

Where the travel to the exit of the protected stairway is along a corridor, its width is determined by the maximum number of persons likely to use the escape route. A corridor 760 mm wide will serve 50 persons; one 1100 mm wide will serve 220 persons; and from 1100 to 1800 mm in 100 mm steps, the number of persons served is one-fifth the number of millimetres width, i.e. a corridor 1500 mm wide will serve 1500/5 = 300 persons.

16.3.3 Compartmentation, fire and smoke stops

Much of the theory and practice of fire control is devoted to containment. In any large building, or one in multiple occupancy, the objective of containments is to restrict the hazard and damage of fire by making sure that it cannot spread from the part of the building where it started. For this purpose, the building is divided up by fire-resisting floors and walls into compartments. The size of these compartments depends on the use made of the building: in a block of flats, each flat is a compartment; in a department store, for instance, the maximum size is either 2000 m^2 floor area or 7000 m^3 volume. Figures for other building uses are given in the Building Regulations.

To be effective, these compartment walls and floors should be imperforate but, in most buildings, this is not possible because of the need to enter the compartment and to run pipes into and out of the enclosed space. Any door ways must be fitted with a self-closing fire-resisting door and any holes for pipes must be fitted or filled with a sealing material to create a fire stop.

Fire stops are also required at the junctions between elements of the structure where there is a risk of smoke or flame passing through. Typical of this is the junction between a brick wall and a concrete roof in which care must be taken to ensure that the roof beams are solidly bedded in a continuous mortar bed; and if precast roof beams of an inverted trough section are used, the fire stopping must follow the line of the soffite (see Fig. 16.2).

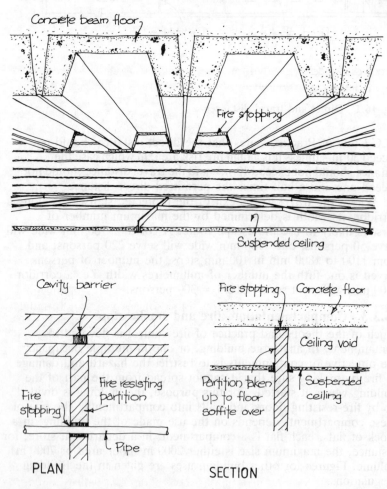

Fig. 16.2 Fire stops and cavity seals

In many buildings there are continuous cavities which, if entered by the fire, provide a ready means for the flames and smoke to spread to other areas. These cavities are usually created by wall linings and suspended ceilings, and must be interrupted at intervals with a fire-resisting barrier. It is particularly important that ceiling void barriers or fire stops are provided along the line of permanent partitions, especially those lining an escape route (see Fig. 16.2).

In any large enclosed space, the main problem following the outbreak of a fire is the spread of smoke. There are more deaths from the effects of smoke than from any other cause in fires. Smoke is a mixture of hot gases given off by the burning material, unburnt or partially burnt particles and air which has been drawn into the fire and heated. It is too hot to breathe and possibly poisonous. It prevents the occupants seeing to find their way out and the fire-fighters finding either the seat of the fire or its extent.

The pattern which develops in a fire is that hot smoke rises rapidly above the seat of the fire and spreads out until it has covered the ceiling. It then begins to descend around the perimeter of the room towards floor level and back towards the seat of the fire. When this smoke layer has reached down to head level, any occupants can be considered to be in extreme danger. This can occur with terrifying speed. Taking the case of a fire in a small workshop of 100 m² floor area, 5 m high, due to the spillage of a highly flammable liquid such as benzene. Assuming that a spark or naked light ignites the liquid to give a flash-fire covering about 3 × 3 m, the time taken for the smoke to fill the room down to 2 m above the floor, i.e. just above head level, would be approximately 14 seconds! And at 20 seconds it would be down to 1.5 m above the floor or shoulder height. It takes about 40 seconds to read this paragraph!

To control this problem, many single-storey factory workshops are now built with screens extending down from the ceiling or, if it is a pitched roof, selected trusses are lined with a non-combustible material. These restrict any lateral spread of hot gases by forming smoke reservoirs. While this restricts the spread, it accelerates the rate at which the depth of the layer of smoke increases. To overcome this problem – and to provide a very satisfactory fire control system – automatic smoke ventilators are fitted in each smoke reservoir (see Fig. 16.3).

16.4 Fire protection

The structural members and the enclosing elements of a building all react differently when subjected to fire and some require protection to ensure their structural stability. Many materials such as metal, stone, glass and concrete, although non-combustible, lose their

SECTION

Fig. 16.3 Control of spread of smoke

strength or integrity in high-temperature conditions; others, such as gypsum and clay products, actually improve in fire conditions, and timber, although combustible, has a high fire endurance if it is of substantial sectional area.

Steel loses rigidity and strength above a temperature of 299 °C (well within the heat from a normal building fire) and consequently buckles and deforms unless protected. Aluminium has an even lower critical temperature and tends to be used just for buildings with a low fire risk. Some natural stones are resistant to the effects of heat but others either expand or shrink and crack or simply decompose. Glass does not require a very high temperature at all before it starts to soften, and unless it is specially treated or strengthened with a wire mesh, it cannot be used in any building element required to provide fire resistance. Concrete disintegrates at high temperatures due either to expansion of the aggregate or to free lime in the cement being converted into quicklime which when, in the presence of water, also expands. In reinforced work, this causes spalling of the surface and exposure of the reinforcement which subsequently rusts, breaking the bond between the steel and the concrete. Careful selection of the aggregate can eliminate failure due to its expansion.

Gypsum products lose their water of crystallisation in heat and in doing so acquire an improved resistance to fire. Clay products are usually formed at temperatures which exceed those encountered in a building fire and, therefore, remain stable.

Timber will burn at first but gradually a layer of charcoal is formed which, in time, prevents any more attack unless the charcoal coating is knocked off. For a timber member to survive a fire, it must be of sufficient girth to accept the loss of its surfaces and still leave sufficient within the member, protected by the charcoal, to carry the loads.

In many building situations, composite constructions are used, one material providing the structural performance and another giving fire protection to the first. The widest application of this principle is structural steel frames. Figure 16.4 shows methods of protecting the steel stanchion and beams of a building frame.

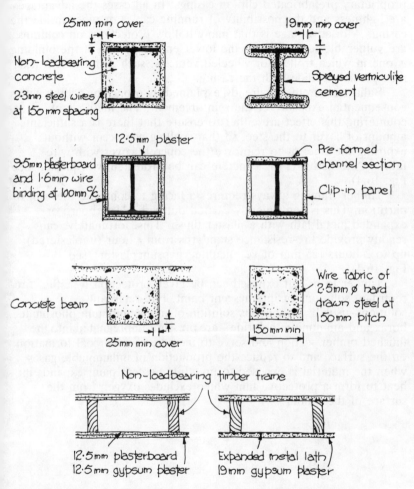

Fig. 16.4 Typical constructions to give one-hour fire protection

Steel protection can be either solid or hollow. Solid protection is achieved by either enclosing the member in concrete or spraying it with vermiculite aggregate and cement. The concrete casing is liable to spall unless the aggregate is carefully chosen; it adds weight to be carried by the beams and stanchions and involves the use of formwork. Vermiculite/cement spray is less expensive to apply but not so robust as concrete and, therefore, more liable to physical damage on stanchions.

Hollow protection can be achieved by timber cradling covered with plasterboard and plaster, plasterboard bound round with wire and plastered, plaster on metal lath or any one of a range of proprietary prefabricated clip-on casings. In all cases the advantages are lightness and the possibility of running cables or pipes inside the casing. A disadvantage is that many hollow protections on columns are vulnerable to damage at the lower end, especially if the building is one in which trolleys or wheeled vehicles, such as fork-lift trucks, may possibly collide with the casing.

Ballast concrete, as already explained, is prone to spalling with consequential exposure of the reinforcement. The methods of countering this effect are either to ensure that there is a sufficient amount of cover to the steel so that spalling can occur without exposing the steel or to reinforce the concrete cover with wire fabric. Alternatively, the concrete can be protected with plaster (see Fig. 16.4).

Timber framing always requires a facing to complete the wall or partition. This is conveniently carried out with plasterboard or expanded metal lath with a plaster finish. This, fortunately, can readily provide fire-resistance standards from $\frac{1}{2}$ hour (unplastered) up to 2 hours (25 mm of vermiculite – gypsum plaster) (see Fig. 16.4).

Timber to be left exposed can be given protection by either fire-retardant solutions or intumescent paint. Fire-retardant solutions–usually an aqueous solution of monammonium phosphate, borax and ammonium chloride – are pressure-impregnated into the finished timber section and serve to increase the charcoal formation on the surface and to reduce the production of inflammable gases when the material is exposed to fire. Intumescent paint expands in heat to form a protective film which excludes oxygen from the surface of the timber.

Chapter 17

The control of commercial and industrial building construction

17.1 Purpose groups

The performance and standards required in a building depend, to a very large extent, on the purpose to which that building is put. It is necessary, therefore, to define the purpose of building types and to set relevant standards. To simplify the range of applications, building functions have been collected in the Building Regulations into eight purpose groups, each representing a number of building functions requiring the same standards. The purpose groups are numbered and given a descriptive title as set out in Table 17.1 below.

This chapter is concerned with the control of aspects of buildings within purpose groups IV, VI and VIII.

17.2 The requirements for structural stability

To achieve stability, the structure of a building must safely sustain the loads of, and applied to, the building and transmit them to the foundations. The foundations, for their part, must safely sustain the loads of the superstructure and transmit them to the ground. In principle, that is all there is to structural stability; in detailed practice, it is much more complex.

The practical application can be considered in four sections: the assessment of the loads to be carried; the structural techniques

176

Table 17.1

Purpose group	Descriptive title	Some typical building types
I	Small residential	Private houses (does not include flats or maisonettes)
II	Institutional	Hospitals Nursing homes Homes for the elderly Schools for the mentally handicapped Residential homes for children under five
III	Other residential	Flats, maisonettes and any other residential buildings not included in Groups I and II
IV	Office	Administrative buildings Clerical offices Drawing offices Computer suites Printing offices Banks and cash offices Telephone exchanges
V	Shop	Retail shops Restaurants Auctioneer's rooms Private lending libraries Hairdressers TV repair shops
VI	Factory	Workshops generally Bottling works Laundries Printing works Film production studios
VII	Other places of assembly	Meeting halls Sports halls Conference rooms
VIII	Storage and general	Goods and material stores Enclosed car parks Commercial garages

employed to carry the loads; the assessment of the bearing capacity of the ground; and the design of the foundations which will be capable of receiving the loads to be supported and which will not overstress the subsoil.

17.2.1 Structural loading

The load to be carried is the total of the self weight of the building, the loads to be placed on the building by the occupants, furnishings and equipment, and the worst situation that could occur of loading due to wind pressure.

The self weight of the building, or dead load, and the imposed loads are calculated in accordance with BS 6399 : Part 1 : 1984, and wind loads in accordance with Part 2 of BS 6399. To calculate the dead load, the volume of each constituent part of the building is found and multiplied by the mass of the material from which it is made. These values are given in BS 648 : 1964 or can be found by testing. Imposed loads present a difficulty in assessment in that in practice they are varied as occupants and equipment move around. They also occur at points of concentration, e.g. a large piece of equipment may occupy much of a floor area but, because it rests on four feet, concentrates its load at four very small points. To overcome this problem, the Code sets out both distributed loads per m^2 and concentrated loads over a 300 mm square for sixty-seven different building types.

Stresses due to either loading must be calculated and the worst case taken as the load on the supporting beams and columns. Flat roofs, i.e. any roof at less than 10° pitch, is treated as a floor and must carry either 1.8 kN concentrated load or a distributed load of 1.5 kN/m^2. Pitched roofs between 10 and 30° must carry 750 N/m^2, reducing to nil when the roof is steeper than 75°.

17.2.2 Structural design techniques

Commercial and industrial buildings may be constructed with load-bearing walls or they may be framed structures of steel, aluminium, reinforced or prestressed concrete, a composite of steel and concrete or, more rarely, timber. Each of these calls for a completely different assessment of structural requirements and capacity and it is recommended that the design should be in compliance with the relevant British Standards and Codes of Practice set out below.

Brick structures	BS 5628 : 1985
Steel structures	BS 449 : Part 2 : 1969
	Addendum No.1 (April 1975) to
	BS 449 : Part 2 : 1969
	Supplement No.1 (PD 3343) to
	BS 449 : Part 1 : 1970
Aluminium structures	CP 118 : 1969
Concrete structures	CP 110 : Part 1 : 1972
	CP 110 : Part 2 : 1972
	CP 110 : Part 3 : 1972
	CP 114 : 1969
	CP 115 : 1969
	CP 116 : 1969
	CP 116 : Addendum No.1 : 1970

Steel and concrete
composite structure   CP 117 : Part 1 : 1965
Timber structure   CP 112 : Part 2 : 1971
                           CP 112 : Part 3 : 1973

17.2.3 Foundations

No direct control exists over the assessment of subsoil bearing
capacity but, in order to design the foundations of the building, a
realistic value for this must be ascertained from examination, site
investigation and laboratory test. The foundation structure itself can
take many forms and still satisfy the simple functional requirement.
In most cases, a satisfactory substructure would result from
following the recommendations of CP 2004 : 1972 or, in the case of
reinforced concrete foundations, CP 110 : Part 1 : 1972 and/or
CP 114 : 1969.

17.3 The requirements for stairways

All stairways, whether they are in places where people live or where
they work, must meet three criteria. They must be ergonomically
efficient, physically safe and comfortable to use. Distinctions are
made, however, between staircases in buildings of different purpose
groups on the basis that a flight of steps in a private house used
regularly by a very small number of people can, because of its
familiarity, be steeper and narrower than one in a public building
which is used by a large number of different people on infrequent
occasions. Ergonomic considerations, i.e. the working efficiency of
the stairs, aim to produce a structure in which the rate of ascent is
commensurate with the effort expended and within the capacity of
everybody using the stairs. To this end, control is exercised over the
minimum going of each step (the horizontal distance from one step
to the next) to ensure that a foot will fit on it; the maximum rise of
each step (the vertical height from one tread to the next) to ensure
that it is within the physical capacity of the average person, and the
combination of the going and the rise to ensure that the flight is
neither too steep nor too shallow. Safety is achieved by defining the
height of handrails, guard rails and balustrades and their capacity to
withstand horizontal loading and by stipulating no fewer than three
steps in a flight so that the change in levels is obvious to the user.
Physical comfort is ensured by requiring a headroom clearance of
2 m and also by limiting the number of treads in a flight to sixteen.

17.4 Requirements for natural lighting

There are no specific regulations requiring a certain standard of
natural light in buildings. Indeed, the only regulations which relate

to window size tend to restrict the amount of natural light reaching the interior of buildings by limiting the size of window openings in the interests of the conservation of fuel and power. There are a number of recommendations for natural lighting and much study has been made of the design of buildings and windows in relation to natural light.

Standards are usually expressed in terms of daylight factor. This is a percentage and represents a comparison between the amount of light falling on a working surface inside the building and the amount falling on a surface outside, i.e. a daylight factor of 2 per cent means that the amount of light reaching, say, a table-top is 2 per cent or 1/50th of the light which it would receive if it was taken outside. Note that this does not give a desirable level of illumination in absolute values as quoted in Chapter 15 for artificial lighting. Nor can such values be given because of the highly variable source from which the light is received. It is also significant that daylight factors, in this country, are based on an overcast sky.

For design purposes, the natural light source is taken as the Standard Overcast Sky which is assumed to give a total illumination of 5000 lux (see Ch. 15 for definition of lux) at ground level. It has been shown that this value, or more, is obtained for 85 per cent of the daylight hours between 8.00 and 17.00.

Daylight factors are recommended as 1 per cent for offices generally, 5 per cent for drawing offices and 5 per cent for top-lit factories. The first two values are related to windows where the light penetrates more or less horizontally, thus illuminating the walls and, by reflectance, the whole room, thereby providing a pleasant working environment. Top lighting tends to leave the walls unlit and gloomy, less light is reflected off the floor and the room seems dim unless a high level of natural lighting is provided. Unfortunately, this high level requires a roof glazing area which is often in conflict with the size restrictions for energy saving. In this case, reliance must be made on artificial lighting.

Two digests by the Building Research Establishment, *Digests 41* and *42*, deal with the estimation of daylight in buildings and the BRE sky component protractors. The latter can be laid over drawings of the building to find the sky component (the amount of light directly entering through the windows) and the external reflected component (the value of the light reflected from other buildings etc). These two values and an internal reflected component (obtained from nomograms, also published by the BRE) are added together to give the daylight factor.

17.5 Requirements for artificial lighting

Chapter 15 explains in detail the terms used, criteria set and methods of calculation of artificial lighting. Table 17.2 gives

180

Table 17.2

Type of space	Amount of light (lux)	Position of measurement
Entrance halls	150	Average at 1.2 m above floor
Corridors	100	Average at 1.2 m above floor
Enquiry desks	500	Desk top
External covered ways	30	Ground
Cloakrooms, toilets	150	Floor
Staff rest rooms	150	Table top
Assembly shops:		
Rough work	300	Working plane
Medium work	500	Working plane
Fine work	1000	Bench top
Very fine work	1500	Bench top
Minute work	3000	Working plane
General offices	500	Desk top
Typing offices	750	Copy
Filing rooms	300	File labels
Conference rooms	750 (with dimming)	Table
Computer rooms	500	Working plane
Drawing office	750	Drawing board
Shops	500	Counter and display

Note: These values may be increased if:
    (a) dangerous or expensive consequences could result from mistakes in perception;
    (b) there are low levels of reflectance or contrast in the job;
    (c) the lit space is windowless.
Source: Based on the Code of the Illuminating Engineering Society.

recommended values for illumination in various working environments. The figures shown give the amount of light, measured in lux, falling on the working plane. The table is based on the Illuminating Engineering Society's Code.

17.6 Requirements for ventilation

Ventilation, as explained in Chapter 15, is required for us to live in comfort and health within our buildings. In the working environment, it becomes even more important because many of the processes of industry produce moisture, fumes, gases or dust, all of which, in excessive quantities, can be injurious to people working in the building. Even where there are no production by-products, the mere presence of a lot of human bodies in one room can require special consideration being given to ventilation standards.
BS 5925 : 1980 points out that the provision of permanent and/or

controllable draught-free ventilation is necessary for all buildings. Designers should take all factors into consideration in determining the ventilation rate necessary for a particular project.

The Standard lists the following situations where there is an absolute necessity for mechanical ventilation to be provided:

(a) rooms or spaces which cannot be ventilated by natural means;
(b) premises where it is essential to remove dust, toxic or noxious contaminants near their source;
(c) hospitals where it is essential to control cross-infection;
(d) where unfavourable external environmental conditions exist such as noise, dust, pollution etc.;
(e) garages or enclosed car parks.

It also lists the following situations where mechanical ventilation is desirable:

(a) factories, in order to remove hot air, moisture and contaminants generally;
(b) domestic bathrooms and kitchens;
(c) assembly halls and lecture theatres where a lot of people are gathered;
(d) tall buildings where wind and stack effect render natural ventilation impracticable;
(e) large commercial kitchens.

Table 17.3 gives recommended rates of fresh air supply to various working environments and is based on BS 5925 : 1980.

Table 17.3 Ventilation rates recommended by the Chartered Institute of Building Services and quoted in BS 5925 : 1980

Type of space	Smoking assumed	Air supply in litres		
		Recommended minimum—whichever greatest		
		per person	per person	per m^2 floor area
Factories	None	8	5	0.8
Open-plan offices	Some	8	5	1.3
Shops and supermarkets	Some	8	5	3.0
Laboratories	Some	12	8	—
Private offices	Heavy	12	8	1.3
Board rooms Executive offices Conference rooms	Very heavy	25	18	6.0
Corridors	—	—	—	1.3
Toilets*	—	—	—	10.0

* Building Regulations: three air changes per hour

17.7 Requirements for space allowances

The Building Regulations, being primarily concerned with the construction rather than the design of buildings, do not impose any relevant standards of space for working people. This is left to two other Acts. The working conditions of people in factories is controlled by the Factories Act of 1961 and that of office and shop employees by the Offices, Shops and Railway Premises Act of 1963. Regulations are gradually being made under the Health and Safety at Work Act of 1974 (which applies to all buildings where people work) and these will, eventually, take over most of the provisions of the two older Acts.

In factories, offices and shops, there is a basic provision of 11.32 m^3 (400 ft^3) per person minimum working space. This figure is further modified in the Factories Act by the rider that any space above 4.27 m (14 ft) from the floor is not to be counted as part of the minimum space. By this means, the Act recognises that, for reasons other than the accommodation of people, factory buildings are often very high. The combined effect of this legislation is that, unless the workshop is less than 14 ft high, each employee must have a minimum of 2.65 m^2 (28.57 ft^2) floor area in which to work. If the ceiling height is less than 14 ft, the floor area must increase to maintain the minimum volume; if the height is more, the minimum floor area figures apply. Therefore, unless other factors dictate, the most economical height for a factory workshop is exactly 4.27 (14 ft).

In an office or shop, the minimum volume is modified by the requirement that there must also be at least 3.72 m^2 (40 ft^2) floor area per person. By the reasoning set out in the last paragraph, it can be seen that the most economical height for an office is 3.05 m (10 ft) exactly. In calculating these figures, it is permitted to ignore furniture, fittings, machinery, plant, equipment and the like. In practice, with a 1500 × 750 mm desk, a filing cabinet, a chair and space to move, the area required is nearly 5 m^2. These standards do not apply to a place to which the public have access, e.g. the sales area of a shop, but the general principle expressed in the Act that the building should not be allowed to become overcrowded must still be observed.

Part C

External works

Chapter 18

Site design

18.1 The siting of buildings and roads

The first task in any design exercise is to decide on the best site layout for the buildings and the roads – if any. The best arrangement for the buildings will be that which allows them to take maximum advantage of the features of the site and its surroundings, and the best arrangement for the roads is that which achieves the purpose for which the road is provided at maximum efficiency in terms of minimum paved area and cost of construction. Over-riding these practical points are the aesthetic considerations of road geometry and the design of groups of buildings. In the case of some building types, the orientation and environment of the site are influential factors in deciding which way to face the buildings to receive the right amount of sun and the most attractive view, but with commercial and industrial buildings the topography or character of the site points of access are likely to be the more important.

18.2 Topography

The word topography means the description of the surface features of any area including not only the land forms, but all other objects and aspects of natural and human origin. The site survey would, therefore, include a topography report on the character of the surface, whether sloping or level, flat or undulating, the natural features such as rivers, streams, ditches and lakes or ponds, the species and size of all trees, any hedges, fences or boundary walls, buildings or structures of any sort above or below ground, and any main service distribution cables or pipes.

A flat level site is the cheapest to develop. Roads can be routed wherever the shortest service direction indicates and building foundations present very little trouble. They also tend to be the least interesting type of development. Any deviation from flat and level brings extra cost and, if used to advantage, extra interest.

Sloping or undulating sites are fairly easy to deal with when the buildings are relatively small, such as housing estates, because the difference in levels between one side of the building and another is not very great. Where, however, the building calls for a large level ground-floor area, such as factory workshop or a sports hall, the difference in levels can be large and expensive earthworks can be incurred.

A notable feature of industrial estates built on a hillside when compared to housing estates in similar positions is the large amount of earth moving that has to be carried out to obtain level areas on which to site the factory buildings, resulting in deep cuttings and embankments. A technique often employed is cut-and-fill in which the excavated material over part of the built area is used to raise the level over the rest of the built area (see Fig. 18.1). The soil uphill from the cut can be either retained by the wall of the building or sloped back, at the angle at which the soil reposes, to form a bank.

Where the building units are small, such as housing, the solution to sloping sites is found in stepping the foundations to follow the

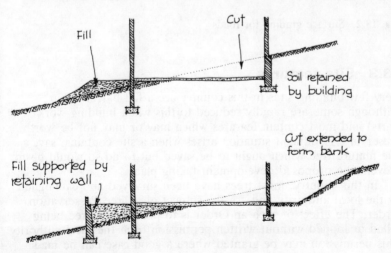

Fig. 18.1 Cut-and-fill

gradient and to provide a suspended ground floor. Alternatively, the problem can be turned to an advantage by a specially designed property which uses the slope by employing a variety of floor levels. The Greeks and Romans demonstrated their understanding of this technique by building their theatres at the bottom of a hill so that the stage was on the plain and the seating in stepped ranks up the hillside.

If the building function calls for a long rectangular plan, then its siting would be arranged so that the long axis of the rectangle ran parallel to the contours thus reducing the amount of cutting and filling required.

Roads can be run either with the contours of the site or across them. If run with the contours, a similar cut-and-fill operation is carried out but, if the slope is steep, there is an added complication in that the buildings one side of the road finish up high above the carriageway while those on the other side are well below it. In either case, vehicular access from the road can be difficult.

Following the site undulations with a road running at right angles to the contours saves on construction costs, but the route requires careful examination to check that the gradients are not too steep nor the changes too sharp. If they are, then the tops of the hills must be lowered and the valleys filled (see Fig. 18.2).

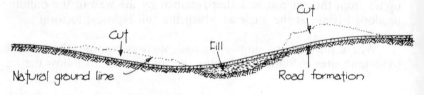

Fig. 18.2 Surface grading for roads

18.3 Site features

Very few building sites in this country are a flat barren plain (although some are rapidly reduced to this when building work starts) and most contain features which may or may not be worth preserving. A difficult situation arises when a site contains, say, a fine mature tree which ought to be saved but to do so would be to prevent the proposed development taking place.

In this country, most trees have been surveyed and catalogued by the local authority and are the subject of a Tree Preservation Order. The effect of such an Order is to prevent the tree being felled or lopped without written permission from the local authority. This permission may be granted where a good case can be made, but it can carry a condition that another tree or several trees must be planted in another part of the site. Even if replacement trees are put in, it takes many years before they can assume the majesty of a fully mature specimen, added to which the developer has the added expense of cutting down the existing tree, grubbing up its roots, buying the replacement trees and planting them.

For these reasons alone, development designers try to avoid the need to disturb existing mature trees but, additionally, the roots left in the ground, even when the tree has gone, will cause subsoil variation which will call for special foundation works.

Foundation design would also be carefully considered if a tree is left and a building is sited near to it, i.e. within a distance equal to its height. Seasonal changes in the quantity of ground water extracted by the roots causes variations in the volume of certain soils and changes in the level of the ground surface by as much as 75 mm between winter and summer. As will be readily appreciated, movement like this is difficult to accommodate in foundation and road design and, therefore, whenever possible, the site layout would be arranged to avoid the problem. Trees are not the only natural feature encountered on sites: many contain ponds, streams or ditches. Any watercourse across an area of land is part of a much larger, finely balanced system of surface water drainage which collects the rain falling on the land, transfers it to our rivers where, eventually, it finds its way to the sea. Disturbance of such a watercourse will inevitably upset this balance, cause flooding upstream and water shortage downstream, unless due care is taken.

Planning the site arrangement to avoid any water is one solution, diverting or piping the stream or ditch is another. In the latter case an extensive survey of the catchment area served by the stream or ditch must be made to ensure that the diversion or pipe size is large enough to take the maximum flow.

The presence of water may be an attractive feature of the site, in which case, not only would it be avoided by the designer, it would also be enhanced by clearance of dead material, if necessary, and provision of measures like paths or benches needed to make it possible to enjoy the facility.

Any site layout where ponds are to be drained and built over requires careful consideration. The subsoil below a pond is, inevitably, different from that found elsewhere on the site. It may be this difference which has caused the pond to be there in the first place. Taking the water away will not, necessarily, affect this difference although it may well alter the pattern of dispersal and level of ground water. In consequence, special precautions must be taken with any substructures, foundations, road beds etc., to overcome any problems that may arise due to differential bearing capacity or subsoil character. What is more, in many cases it will be necessary to do a certain amount of filling which will futher affect any substructures since the load-bearing elements must penetrate this filling to reach a more satisfactory, and stable, established lower strata.

Besides natural features, many development sites have been the subject of earlier building works, leaving a legacy of useful structures or buildings which are not required, walls, fences and similar artificial features. Where existing buildings are to be retained and, possibly, adapted, the layout of the rest of the site will be affected considerably. Roads must be laid to serve these buildings and these, in turn, will dictate where the other buildings in the development must be sited.

The demolition of unwanted buildings and walls does not, usually, present much difficulty and effectively their presence can be ignored when designing the site, but due regard must be paid to the problems that old foundations and, worse still, old basements can create for structural and highway engineers trying to work over them.

18.4 Subsoil

Foundation systems for buildings and road construction systems exist which can solve the problems posed by any type of subsoil, but the fact that any problems are solvable does not mean that in laying out a site the designer would not try to avoid the problem in the first place. Any areas of the site which are likely to need complex and, therefore, expensive foundations may be avoided altogether – if they are small in extent – and retained as recreation areas or paved as a car park. Where the density of development does not allow for any of the site to remain unused or the poor ground is extensive in area, small, low-rise or single-storey lightly-loaded buildings, such as garages, stores, toilet blocks etc., may be placed in these parts of the site and the additional foundation costs accepted as the price of high-density development.

The situation the designer would be at pains to avoid is placing a building so that it straddles a significant change in the subsoil. All buildings settle to some extent when first erected, due to compression of the subsoil; different subsoil compress at different rates and to a variable extent. If the movement is uniform over the whole of the site covered by the building, no damage will ensue but, where there is a change of subsoil, the settlement will not be uniform. Some parts may settle quickly and by only a small amount; others may move gradually over a period of time but to a greater extent. This would be the case if part of the building was founded on gravel and part on clay. Such differential movement will inevitably lead to rupturing of the building fabric. After the initial settlement, different soils respond in differing ways to climatic changes with, in the circumstances being considered, further damage to the building. These effects cannot be prevented and, therefore, must be avoided by siting each building so that its foundations all rest on similar subsoil.

18.5 Access

In any building development, provision must be made for pedestrians, and goods to enter and leave the site. These points of entry and exit are determined, not by the desires of the designer or

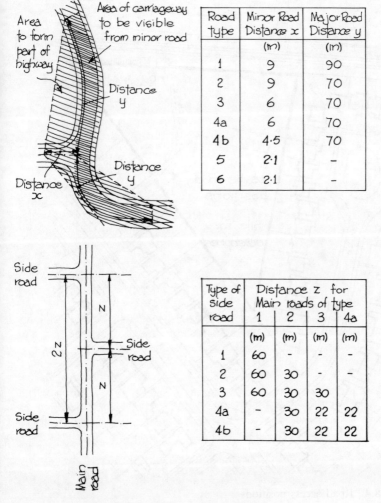

Road type	Minor Road Distance x	Major Road Distance y
	(m)	(m)
1	9	90
2	9	70
3	6	70
4a	6	70
4b	4·5	70
5	2·1	–
6	2·1	

Type of Side road	Distance z for Main roads of type			
	1	2	3	4a
	(m)	(m)	(m)	(m)
1	60	–	–	–
2	60	30	–	–
3	60	30	30	
4a	–	30	22	22
4b	–	30	22	22

Fig. 18.3 Road junction sight lines and staggers

190
190

the characteristics of the site, but by the location and character of the existing roads to which the new are to be connected.

18.5.1 Vehicular access

With vehicular access to the site, the new roads must, of course, connect to one or more existing roads adjoining the site. Even when there is a road conveniently running the length of one side of the site, the exact point at which a junction or junctions can be made will be controlled to take account of the hazard that can be created by vehicles turning into, or emerging from, the estate. This is

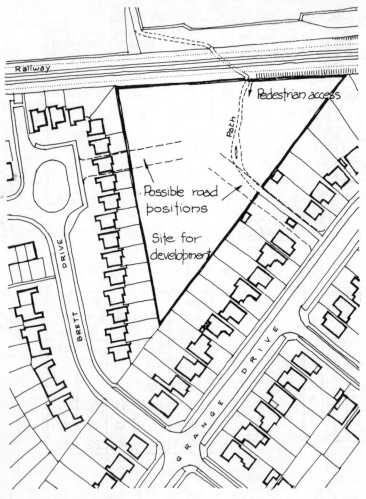

Fig. 18.4 Road access positions

Table 18.1

Type of road	Function	Width of carriageway (m)	Garages and hardstandings served (max)	Design speed km/h	Design speed (mph)	Access only from
1 Local distributor	Link between district distributors and access roads	7.3	No frontage access	50	(30)	Primary distributor District distributor Road type 1
2 Major access	Direct access to buildings and land	6.75	No limit*	30	(20)	Road types 1 and 2
3 Intermediate access	Minor loop roads Culs-de-sac over 300 m long	6.0	400	30	(20)	Road types 1, 2 and 3
4A Minor access	Minor loop roads Culs-de-sac not exceeding 300 m	5.5	200	15	(10)	Road types 2, 3 and 4A
4B Minor access	Minor loop roads Culs-de-sac not exceeding 100 m	5.5	75	15	(10)	Road types 2, 3, 4A and 4B
5 Mews Court	Culs-de-sac not exceeding 36 m	5.5**	25	Very low		Road types 3, 4A and 4B
6 Private drive	Access to dwellings	2.5 for single dwellings 4.25 for shared drive	3	Very low		Road types 2*, 3, 4A, 4B and 5

* Turning facilities must be provided within the house boundaries so that cars can be driven off and on to a type 2 road in a forward gear.
** Combined vehicle and pedestrian access–width between entrance walls.
Source: Based on Table 4, *Design Guide for Residential Areas*–Essex County Council

particularly critical if there are any other junctions or bends in the existing road (see Fig. 18.3).

In many cases, the existing road layout or the presence of buildings between the road and the site eliminates all but one possible point of entry (see Fig. 18.4).

When considering the layout, due regard must be paid to the hierarchy of vehicle routes. This, as laid down in the *Design Guide for Residential Areas* by Essex County Council, consists of seven different types of road within a housing estate with, in addition, primary distributor and district distributor roads serving national and district purposes. Because of their differing functions and design speeds, certain minor roads are not allowed to gain direct access from the more major roads. The seven road types, their width,

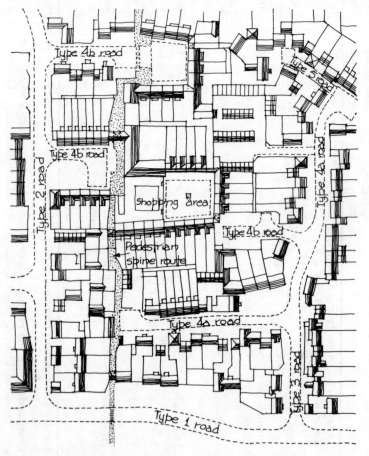

Fig. 18.5 Hierarchy of roads and pedestrian spine routes based on the *Design Guide for Residential Areas* by Essex County Council

design speed and the restrictions to type of road from which they may have access are shown in Table 18.1 and illustrated in Fig. 18.5.

18.5.2 Pedestrian access

In many estate layouts, pedestrian access is gained by footpaths laid alongside the roads. Where this applies, the pedestrian must be protected from the traffic, as far as possible, by the provision of vertical kerbing, which presents a barrier to vehicles, and a grass verge which separates the road and the footpath.

Even with these provisions, dangers still exist for pedestrians, especially young ones. Furthermore, the point of access allowed for vehicles may not be the most convenient point for pedestrians. Footpaths, of course, do not present the same troubles as roads in relation to accidents due to speed or lack of visibility and, therefore, can connect to the existing road at any convenient point. Alternatively, they can lead directly to other pedestrian areas such as shopping precincts. With these considerations in mind it can be seen that two or more points of access to the site, for pedestrians only, may be provided connncting to a layout of pedestrian spine routes, leading directly to the houses and divorced from the roads (see Fig. 18.5).

Chapter 19

Excavations and changes of level

19.1 Problems encountered

Whenever a hole or trench is excavated into a site, two problems can be anticipated: the need to prevent the sides of the excavation from falling into the hole or trench, and the need to keep the excavation clear of water. Like most principles in building construction, this is not an invariable rule. Excavations in certain soils do not require any support if they fulfil certain conditions and a few subsoil operations can be carried out with the excavation partly full of water. The laying of underground pipes, for instance, can continue even if the trench has water in it, as long as the pipe is of the continuous plastic type which is simply laid in the bottom of a level trench. Mass concrete can be placed under water through a tube called a tremie pipe, and will set satisfactorily, but the water must be still and the operation requires somebody with experience of this type of work to be in charge. However, in many cases, the sides of the excavation will not stand on their own for the time required. In addition, other factors such as heavy vehicles running on the surface near the trench can make temporary support of the sides a necessity and, unless the site operation is one of the types such as are described above, the intruding ground and surface water will have to be removed.

Where changes in the level of the surface are created, the problems are much the same as for excavations, i.e. support and ground water.

Support is only required if the angle of bank between the higher and lower levels is greater than the natural angle of repose of the soil (see section 19.4). Below this angle, a natural bank can be formed. In many cases the 'step' from one level to the other is required to be vertical. To achieve this, a permanent retaining wall would be built.

Ground water problems are partly the same as for excavations, i.e. the working area should be kept dry, but there is also the permanent problem of water pressure on the back of the retaining wall.

19.2 The need for temporary support of excavations

The need for temporary support arises from the following factors either singly or in combination:

(a) the type of subsoil;
(b) the depth of the excavation;
(c) the presence of ground water;
(d) the duration of the work;
(e) any surcharges placed adjacent to the excavation;
(f) any activities which may generate vibrations in the ground.

19.2.1 Subsoil types

Natural sobsoils generally fall into cohesive or non-cohesive classes. Cohesive soils are those in which the particles cling together, such as clay. Non-cohesive soils are granular and 'flow' when unconfined: sand is a non-cohesive soil. Nature being what it is, a combination of both types is more commonly found than pure specimens of either and, therefore, in a site investigation report, one finds descriptions of sandy clay or clayey sand as well as sandy gravel and boulder clay.

Generally, the non-cohesive soils offer more reliable foundations than the cohesive soils but are more prone to trench collapse. In both cases, drying of the exposed trench face can alter the characteristics of the initially excavated material to a significant extent.

Excavations also have to be taken down through subsoils that are not natural. These may be areas where the level has been raised or filled up – referred to as 'made ground' – or it may be only in part of the excavation where it crosses earlier earthworks which have been back-filled. Both made ground and back filling are subsoil conditions calling for careful examination in connection with the subject of temporary support.

19.2.2 Excavation depth

The Construction (General Provisions) Regulations 1961 consider
4 ft (1.219 m) the critical depth; any excavation in excess of this in
which there is a danger of material collapsing or falling in must be
supported.

In this context it should be borne in mind that the mass of only
a small quantity of soil falling on to men working in a trench can
cause their deaths by preventing the expansion of their chests and
lungs, thereby causing suffocation. (Subsoil has a density between
about 1500 and about 2000 kg/m^3.) If, however, the trench is only
just over a metre deep, the quantity of material which could
collapse inwards is restricted and, furthermore, it is not very likely
that such a collapse would bury the upper torso of men working in
the trench.

Beyond a depth of 1.2 m, decisions on the type and extent of
trench support will be influenced by all the other factors listed in
section 19.2.

19.2.3 Ground water

The presence of water in the ground into which an excavation is
taken creates difficult working conditions and endangers the stability
of the sides of the excavation.

There are two main effects of ground water. Firstly, a flow
entering through the sides of the trench and softening, loosening
and carrying with it the subsoil of the trench face. This undermines
the upper dry layers, leading to collapse. Secondly, 'blowing' or
'boiling', which is an upsurge of water through the bottom of the
excavation, particularly where the sides have been lined with steel
sheet piling to exclude the water. Control of this subsoil water is
dealt with in section 19.5.

19.2.4 Duration of the work

If the building operations are only going to take an hour or two, the
necessity to support the sides of the trench is diminished,
particularly in cohesive soils. But, if the excavation is to remain
open for a long period, swelling and cracking of the soil can occur,
leading to a loss of stability and dangerous conditions. The swelling
is due to the removal of the lateral pressure which formerly was
imposed by the material now excavated. This lack of restraining
pressure allows the trench side to move inwards and, consequently,
cracks develop in the surface. Further cracking occurs due to drying
of the subsoil. In wet weather, these cracks can fill with water
causing a softening of the subsoil and a serious risk of collapse (see
Fig. 19.1).

19.2.5 Surcharges

The subsoil in danger of collapsing into the excavation is that part

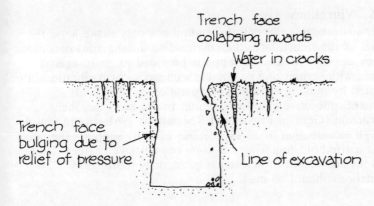

Fig. 19.1 Collapse of trenches in clay

contained within a triangle bounded by the trench face, the ground surface and the angle of repose. Angle of repose is explained in more detail in section 19.4 and is also known as the angle of internal friction. It is the strength of this internal friction or sticking together of the soil particles which determines whether the triangle of soil will rupture along the angle of repose and slide down into the trench. If the downward force of the triangle is increased, the resistance to rupture may be overcome and a collapse occurs. This force can be increased by placing a load on the ground surface between the edge of the trench and the end of the line of the angle of repose. Such loading can result from materials being deposited or stacked in this position or by a heavily-loaded vehicle being driven too close to the trench (see Fig. 19.2).

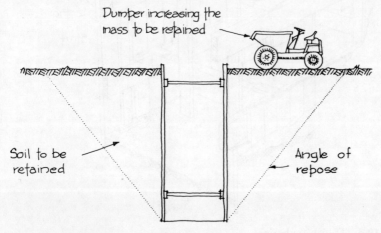

Fig. 19.2 Surcharge on retained subsoil

198

19.2.6 Vibrations

Vibrations travelling through the ground not only shake loose the material of the trench face, thereby creating a dangerous condition, but they can also loosen the supports provided to guard against collapse, with serious and deadly consequences. Such vibrations are generated by moving vehicles, the operation of plant and equipment, pile driving and shot firing. In the last case, the Construction (General Provisions) Regulations 1961 require that a thorough examination of all excavations and any supports they may have is carried out following the firing of explosive charges. If other vibrations are known to be occurring, regular, thorough examinations should be made.

19.3 Methods of temporary support of excavations

The simplest and most widely used method of support for the normal shallow excavation is that of trench timbering. In the past this has been carried out using timber poling boards, walings and struts (see Fig. 19.3). Present practice tends to favour steel trench

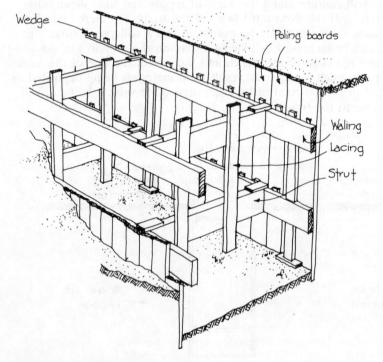

Fig. 19.3 Trench timbering

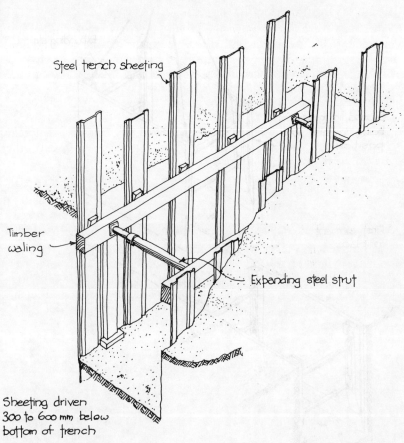

Steel trench sheeting

Timber waling

Expanding steel strut

Sheeting driven
300 to 600 mm below
bottom of trench

Fig. 19.4 Steel trench support

sheeting and adjustable steel props (see Fig. 19.4). There are also methods using concrete sheet piling and proprietary, all-steel trench supports which are lowered bodily into the trench and then expanded by the action of hydraulic rams to apply pressure to the trench walls (see Fig. 19.5).

Lining the trench after completion or while it is being excavated requires the subsoil to be sufficiently cohesive to hold up by itself until the timbering is in place. In situations of very loose or wet soils, this requirement may not be met and a support must be provided before excavation can commence. For shallow trenches and pits, this is usually achieved by driving sheet or concrete piling (more rarely timber is used) down to the required level along the lines of the excavation before the digging commences. For deep working in bad ground, a more advanced technique using cofferdams of caissons or a dewatering method must be employed.

200

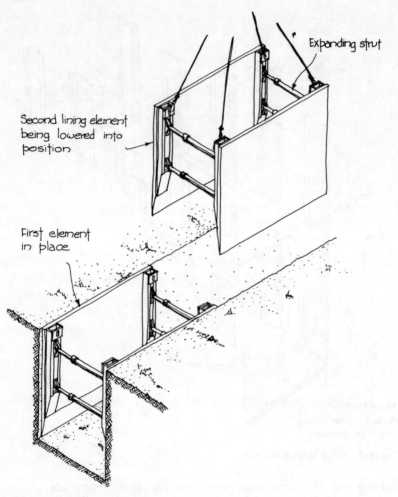

Expanding strut

Second lining element being lowered into position

First element in place

Fig. 19.5 Proprietary trench support

19.4 Retaining walls

The permanent support of earth faces is achieved by a variety of retaining walls. These may be of brick, precast concrete units, *in-situ* concrete structures or prestressed concrete. They may be provided where there is a change in land of the site surface or they may be the enclosing walls of a basement. Reinforced and prestressed concrete retaining walls and basement constructions are the subject of more advanced study (see *Advanced Building Construction*, Vol 1 and 2). This section will deal with the subject of mass retaining walls of brick or concrete block. To ensure that a

201

wall will retain the soil behind it, the characteristics of the retained material and the mass to be supported must be studied. All dry granular materials assume a conical shape when tipped on to a flat surface, and the angle of the surface of the cone remains the same no matter how high the pile. This angle is known as the angle of repose (see Fig. 19.6) and is affected by the presence of moisture.

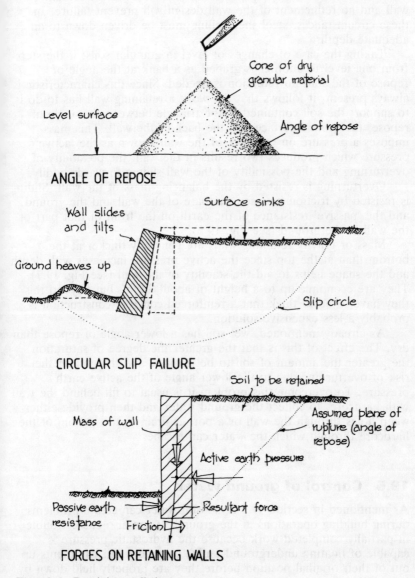

Level surface

Cone of dry
granular material

Angle of repose

ANGLE OF REPOSE

Wall slides
and tilts

Surface sinks

Ground heaves up

Slip circle

CIRCULAR SLIP FAILURE

Soil to be retained

Mass of wall

Assumed plane of
rupture (angle of
repose)

Active earth pressure

Passive earth
resistance

Friction

Resultant force

FORCES ON RETAINING WALLS

Fig. 19.6 Retaining wall theory

Damp soils will stand at a steeper angle than dry ones, but saturated soils flow more easily and will have a much lower angle or even no angle at all if they are very wet.

Clay soils do not have an angle of repose because of their cohesive nature, but exhibit a tendency to rupture and slide along a circular slip plane (see Fig. 19.6). As can be seen from the illustration, the circular slip plane can pass beneath the retaining wall and no refinement of the wall design will prevent failure. In these circumstances, steel sheet piling must be driven down to an adequate depth.

Taking the case of changes of level in granular soils; if the step from one level to another is graded as a bank at the angle of repose of the soil, no retention is needed. Since this characteristic is always present, it follows that the work a retaining wall has to do is to support the soil contained in the triangle between the angle of repose, the ground surface and the back of the wall. This mass imposes a pressure on the back of the wall known as the 'active' pressure which produces two results in the wall: the possibility of overturning and the possibility of the wall sliding forwards bodily.

Overturning is resisted by the balancing mass of the wall. Sliding is resisted by friction between the base of the wall and the ground, and the 'passive' resistance of the earth on the front of that part of the wall in the ground (see Fig. 19.6).

Mass or gravity retaining walls are generally thicker at the bottom than at the top since the active pressure increases with depth and the shape helps to aid the stability of the wall (see Fig. 19.7). They are economic up to a height of about 2.0 m, but beyond this they have to be so thick that a reinforced concrete construction is probably a less expensive solution.

As already mentioned, wet soil has a lower angle of repose than dry. The effect of this is that the greater the degree of saturation the greater the amount of soil to be retained and the greater the risk of overturning due to the lower angle of the active earth pressure. To overcome this problem it is usual to fill behind the wall with hardcore, to collect the ground water and then provide either weep holes through the wall or a porous drain at the bottom of the hardcore through which the water can escape.

19.5 Control of ground water

As mentioned in section 19.2.3, subsoil water can create problems during building operations in the ground. It can also cause trouble in partially completed work because the hydrostatic pressure is capable of floating underground tanks, chambers and basements up out of their original position before they are properly held down by being filled (in the case of tanks) or by the load of the complete

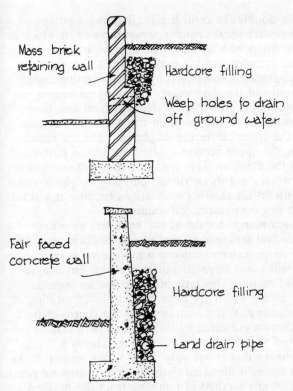

Mass brick retaining wall

Hardcore filling

Weep holes to drain off ground water

Fair faced concrete wall

Hardcore filling

Land drain pipe

Fig. 19.7 Retaining walls

superstructure (in the case of basements). In both cases, control of ground and surface water and possibly lowering of the water table must be maintained until the danger period is passed.

Surface water, i.e. rainwater running across the surface of the site, following the natural slope and entering any excavations, can be collected in a shallow trench dug uphill from the area of operations and either pumped or drained away.

Subsoil water can be dealt with by either allowing it to enter the excavation and then pumping it out or preventing it from entering at all.

19.5.1 Pumping excavations

This is the simplest, most direct and commonly used method of dealing with ground water and consists of a sump dug below the lowest level of the excavation into which the lift pipe of a suitable pump is inserted. The sump is to collect the water and is sometimes assisted in this by a small grip or ditch dug along the excavation, falling towards the sump.

Several types of pump are in use but are mainly either

diaphragm (single or double) or centrifugal. Diaphragm pumps consist of a large-diameter short cylinder across the end of which is a flexible diaphragm fitted with a valve. The inlet pipe is connected to the bottom of the cylinder and the outlet is fitted above the diaphragm. The equipment is completed by a reciprocating arm connected to the centre of the diaphragm, driven by an engine or motor. As the diaphragm is pulled up, water is sucked into the cylinder, on the downward stroke; some of the water passes through the valve and into the space above the diaphragm. On the return stroke, the water in the upper section is lifted out of the outlet and more is drawn into the cylinder. This type of pump will not handle great quantities of water, and its maximum pumping height is about 6 m, but it is very useful for clearing excavations because it will take water with up to 15 per cent suspended solids.

Double-diaphragm pumps consist of two cylinders mounted alongside each other and arranged so that as one is discharged, the other is filling with more water to achieve a more even flow. Centrifugal pumps will move large quantities of water but it must not contain any solid matter. This type of pump has an impeller, consisting of a disc across which are fixed curved blades rotating at high speed within a casing. Water is drawn into the centre of the impeller and then thrown outwards by the blades where it is collected and directed to the outlet by the casing. This is a lift and force pump which means that it not only draws water up out of the excavation but also forces it along an outlet pipe to a disposal point.

A disadvantage of any method of pumping from the finished excavation is that the ground water is induced to flow towards the excavation, thus increasing the risk of collapse.

19.5.2 Garland drain

On sloping sites or sites where the excavation passes through a permeable strata and into a clay bed, it may be advantageous to collect the surface or subsoil water before it reaches the trench or pit by means of a garland drain. A garland drain consists of a shallow trench or channel arranged to intercept the water flow and direct it into a disposal point or away from the site of the building.

Surface water collection, usually only necessary in clay soils, can be effected by a garland drain trench dug slightly across the contours of the site to obtain the necessary fall, connecting to a convenient ditch, watercourse or surface water sewer.

Subsoil water flow often occurs at the level where an impermeable lower strata is overlaid by a permeable upper strata. It is at this point where a garland drain trench or channel would be formed (see Fig. 19.8).

19.5.3 Well point dewatering

When simple pumping from a sump or a garland drain are

205

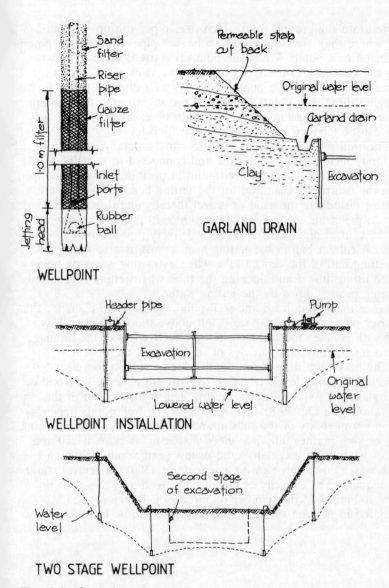

WELLPOINT

GARLAND DRAIN

WELLPOINT INSTALLATION

TWO STAGE WELLPOINT

Fig. 19.8 Ground water control

inadequate to deal with the ground water situation, more effective methods must be adopted. Some of these, such as freezing the ground, are expensive and only resorted to on very large complex projects, but the well point system is one which is available for the more normal size or type of contract in wet non-cohesive ground.

It consists of as many as fifty to sixty filter wells or well points inserted into the ground from which riser pipes run up to ground-level where they are connected to a header pipe. This header pipe is connected to a pump. When the pump is started, water is drawn from the ground into the well point, up the riser pipe, along the header pipe, through the pump and away to a disposal point. By arranging these well points either in a rectangle round the building site or alongside the line of the trench to be excavated, the water table can be lowered to below the excavation level.

The well point consists of a 60–75 mm diameter gauze screen 1 m long, enclosing a central tube and connected to a jetting head (see Fig. 19.8). To place the well point in position, water is pumped down the central tube and out of the jetting head, which has the effect of dislodging the sand or gravel thereby allowing the well point to penetrate the soil with ease. Once in position, jetting stops and the central pipe then serves as a delivery pipe to the extract pump. A rubber ball in the jetting head serves as a valve, closing the jetting end of the central tube when extracting water by being sucked up against it and opening the tube when jetting is required by being pushed down by the jetting water.

Since the jetting head and well point are larger in diameter than the riser pipe an annular space is left above the well point. This is filled with coarse sand to increase the effectiveness of the drainage and to restrict any fine particles of soil entering the system.

The maximum draw-down depth which can be achieved is 5.0 – 5.5 m below pump level. If greater depths are needed, a second or more stages of well points can be installed in the bottom of the excavation (see Fig. 19.8). The spacing of the well points depends on the permeability of the subsoil and the quantity of water flowing. Soils of high permeability may need a spacing as close as 300 mm centres whereas in, say, silty sand of low permeability (through which water flows only slowly) a spacing of 1500 mm centres could be adequate. The capacity of a well point with a 50 mm riser is about 10 l/min and the number required can, therefore, be calculated in relation to the quantity of water to be handled.

Chapter 20

Estate roads and footpaths

20.1 Road structures

Figure 18.5 and Table 18.1 show the types of road which might be constructed by a builder. Primary and district distributor roads, i.e. the main trunk roads, are usually constructed by a civil engineering organisation and their technology is outside the scope of this chapter.

There are three types of road construction used in the formation of estate roads, rigid, flexible or block. The first two may be used for any of the classes of road shown in Table 18.1. The last tends to be confined to roads designed for low speeds and private drives, although design speeds up to 50 km/h are claimed to be acceptable making block construction suitable for all types of estate road. The terms rigid and flexible suggest that one form of road stays absolutely flat in use while the other bends. This is not strictly true, but the rigid construction is more prone to cracking than the flexible one, while a flexible pavement is considered not to possess any tensile strength. In their joint publication, *A Guide to Concrete Road Construction*, The Department of Transport, Transport and Road Research Laboratory and the Cement and Concrete Association define the functions a road must fulfil as:

1. To carry the wheel loads applied by traffic and to distribute the resultant loading over the soil beneath so that the stress induced in the soil does not exceed a safe value for the sub-grade.

2. To provide a running surface that will be safe and sufficiently strong to resist the effects of traffic and weather, particularly frost, with a minimum of maintenance over a long period.

3. To protect the underlying soil from an ingress of water and consequent variations in moisture content and thus assist in the maintenance, during the life of the road, of a uniform state of stability in the soil.

The wheel loads referred to in (1) are both static and dynamic, i.e. standing and moving loads, each of which induces a different form and extent of stress in the road structure. The construction of a road with low maintenance costs over a long period, no matter what the traffic or weather conditions, has obvious attractions for whoever is to own the road. Frost, as mentioned, is particularly aggressive to roads: water can seep down into minor cracks and gaps in joints and then expand as it freezes to fracture roads dislodge kerbs and footpaths. The third function, the maintenance of stability in the ground, is particularly relevant where the subsoil is clay, or a material similarly susceptible to changes following variations in the weather.

20.2 Terminology

There are a number of terms used in connection with the design and construction of roads. The word to be particularly noted is 'pavement' which, in this context, means the whole structure of the road or carriageway laid on the soil, not, as popularly used, the footpath alongside the carriageway. The following list and Fig. 20.1 give a selection of the rest of the terms used:

Sub-grade: The upper part of the soil which supports the pavement.

Formation: The surface of the sub-grade in its final shape and levels after completion of the earthworks.

Sub-base: The material between the sub-grade and the slab or roadbase.

Roadbase: The layer which supports the surfacing.

Slab: Plain or reinforced concrete which may be just the roadbase or may be roadbase and running surface combined.

Surface: The top layer which provides the wearing coat on which the traffic runs.

20.3 Rigid road pavements

As explained above, rigid pavements comprise a concrete slab, plain or reinforced, laid over a sub-base. The upper face of the concrete

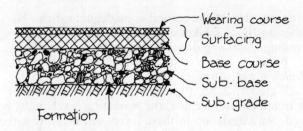

FLEXIBLE PAVEMENT

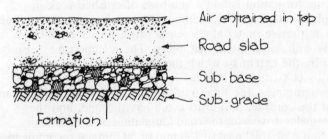

RIGID PAVEMENT

Fig. 20.1 Road terminology

may be the finished running surface or it may be overlaid with a bituminous wearing course in one or two layers.

No matter how well the pavement is constructed, its ability to fulfil its function is dependent upon the support it obtains from the sub-grade. The construction selected must match the strength of the sub-grade which depends on the nature of the soil, its plasticity and its dampness or depth of water table. These, when taken together, affect the Californian Bearing Ratio (CBR). This ratio expresses the comparison between the load required to produce a standard penetration – either 2.5 or 5.0 mm – of a hardened steel plunger into a sample of the soil and the load required to produce the same penetration in a 'standard' crushed rock sample. The test, set out in BS 1377 : 1975, was originally developed in California to assess the quality of crushed rock for roadbase material. It is now adopted as a method of assessing the strength of the natural sub-grade. A normal sub-grade is defined in *A guide to the structural design of pavements for new roads*, Department of the Environment, Road Note 29, as being one in which the CBR value is between 2 and 15 per cent. Where the value is less than 2 per cent, the slab should be increased by 25 mm over the recommended thickness, and where it is more than 15 per cent, a 25 mm reduction in slab thinckness can be made.

If the bearing capacity of the subsoil is inadequate it can be increased by draining (if its weakness is due to excess water), compacted by rolling, ramming or vibrating, or by soil stabilisation using cement or bitumen. These techniques are dealt with in detail in *Advanced Building Construction, Vol. 2*.

When the sub-grade has been levelled and rolled or given whatever other treatment is specified, the formation levels must be carefully checked. Any deviation in these levels will mean that the road construction above formation will be less than specified and, therefore, weaker; or more than specified and, therefore, more expensive than was allowed.

Over the formation is laid a sub-base of crushed rock or concrete plus sand or gravel, lean concrete or a mixture of soil and cement. The precise material chosen depends on its availability and, particularly with soil/cement, suitability. The selection may also be influenced by the extent to which the sub-base is to be used as a haul road by construction vehicles, especially in winter. In most building programmes, the final finish of the road is not applied until the end of the contract period to avoid any damage, and the builder's vehicles drive over the road foundation construction or sub-base. Sub-base thickness of 150 mm or 80 mm are recommended for 'weak' or 'normal' sub-grades, but these may need to be increased as shown in Table 20.1 to support construction traffic.

Table 20.1 Sub-base thicknesses for concrete pavements

CBR of sub-grade (per cent)	Sub-base thickness (mm)
Less than 2	280
2 – 4	180
4 – 6	130
6 – 15	80
over 15	0

Before laying the concrete slab, pressed steel road forms are positioned and fixed with steel stakes to contain the edges of the concrete and to give the finished level of the top surface. Between the road forms and over the sub-base is laid polythene sheet to provide a slip membrane which will allow the concrete to move slightly with weather changes, and this will also prevent loss of cement solution from the wet mix (see Fig. 20.2).

The concrete used is generally specified to have a strength of 28 MN/m^2 at 28 days with a water/cement ratio of not more than 0.5 by weight. It is also recommended that air entraining agents should be incorporated in the top 50 mm to produce between 3 and 6 per

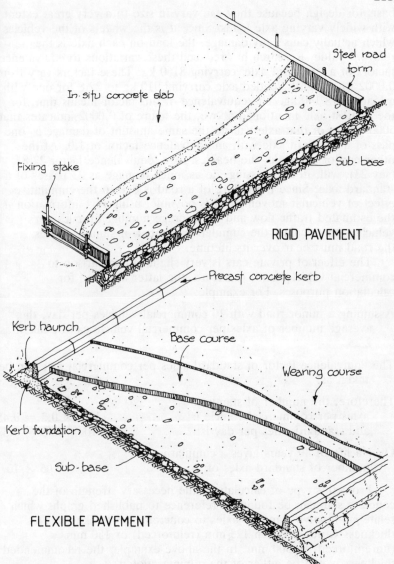

Fig. 20.2 Typical road constructions

cent of minute bubbles in the concrete. This has the effect to preventing the slab becoming saturated by capillary action.

The thickness of the slab depends on the density of the anticipated traffic and the anticipated damage the constant passage of this traffic can cause. Merely counting the number of vehicles which could be expected to pass a given point is not an adequate basis for design because they can vary in size to a very great extent with widely varying axle loads. Since it is the wheels of the vehicles which actually cause any damage, the load on each axle is the relevant value. Research has reduced these variations to equivalence factors of a 'standard axle' carrying 8160 kg. These factors vary from 0.0002 for a lightly loaded axle carrying 910 kg to 22.8 for one with a load of 18 140 kg. The equivalence of this factor means that, for the lightest axle mentioned above, the figure of 0.0002 indicates that 5000 passes of that axle will do the same amount of damage as one pass of a standard axle with an equivalence factor of 1.0. At the other end of the scale, the heaviest axle, equivalence factor 22.8 (say 23), will, in one passing, do as much damage as 23 passes of a standard axle. Since the failure of a road is due to the cumulative effect of vehicular movement it is possible, using this information – the estimated traffic flow and the average number of axles per vehicle – to calculate the cumulative number of standard axles the road must bear over its lifetime.

The effect of private cars is very slight in comparison to commercial vehicles and, therefore, the latter are used for calculation purposes. For example:

Assuming a minor road with 10 commercial vehicles per day, the average number of axles per commercial vehicle can be taken as 2.25.

The equivalence factor of standard axles per commercial axle can be taken as: 0.2

Therefore, the number of standard
axles per commercial vehicle is: $2.25 \times 0.2 = 0.45$
and standard axles per day is: $10 \times 0.45 = 4.5$

Over a life of 40 years gives a cumulative
number of standard axles of: 0.06×10^6

From this type of calculation, the necessary strength of the concrete slab can be found by reference to published graphs which relate cumulative standard axles to concrete slab thickness. This thickness varies between 125 mm (reinforced) or 150 mm (unreinforced) to 300 mm. In the above example, the recommended thickness would be either of the minima quoted.

20.4 Flexible road pavements

Flexible pavements are the tarmacadam type of road, consisting of a sub-base, a base course and a wearing course, and gravelled access pavements and drives consisting of a sub-base and a screeded gravel surfacing.

As with rigid pavements, correct preparation of the sub-grade is essential to produce a durable and economic construction. The soil type and bearing capacity must be assessed and improved as necessary to give a satisfactory CBR and the surface must be worked to the required formation levels.

The sub-base is laid directly on to the formation and consists of materials similar to those used for a concrete road – usually crushed rock or lean-mix concrete mixed in the proportion of one part cement to fifteen parts all-in aggregate laid dry. The selected material is laid in layers between 100 and 150 mm thick and thoroughly compacted by rolling. The precise thickness required is determined by the CBR of the sub-grade and the cumulative number of standard axles, and can vary between 80 and 600 mm.

In estate development using flexible pavements, the sub-base is always used as the road by the builder's vehicles during the construction period. Such use as a haul road can cause hollows to form in the surface of the sub-base, which subsequently must be filled before the base course can be laid, but it does ensure that the road foundation is well consolidated and it is also cheaper to do this filling than it would be to repair the macadam surface if the road had been completed at an early stage in the contract.

If the upper course of the road is to be a bound surface (the paving material is bound together with bitumen or a similar tar product), a base course of rolled asphalt, tarmacadam or bitumen macadam is laid on the sub-base, followed by a wearing course of hot rolled asphalt, bitumen macadam, tar surfacing or cold asphalt. The thickness of the base course should not be less than 60 mm and can be more if its surface is worked to the necessary falls and gradients. The wearing course is usually 20 to 40 mm thick (see Fig. 20.2).

A gravel finish gives an unbound surface which is much cheaper than a bound surface, but is only suitable for areas where speeds are very low and hard braking and acceleration are not anticipated. It is, therefore, restricted to private domestic drives for which it is quite popular, not only because of its low cost, but also because the householder can be forewarned of an approaching vehicle or pedestrian by the crunching of the gravel – a feature much appreciated in isolated country properties.

Before laying the gravel, the sub-base can be sprayed with tar to seal the surface, thereby discouraging the growth of weeds and grass. The topping is graded to pass a 40 mm ring with enough sand to bind the gravel together. It is laid dry and rolled to give a finished thickness between 25 and 50 mm. Although it appears pervious, the surface voids can fill with small particles to become resistant to the passage of water and, therefore, it must be laid to falls. It is also necessary to restore the surface periodically by raking the gravel back to the original levels and treating it with a total weed-killer.

20.5 Block roads

The use of blocks as a road construction is of ancient origin: the Roman roads were built this way and the principle has been employed continuously ever since. Originally the blocks were natural stone cobbles or setts, but today we use a precast concrete interlocking block. The modern block gives a much better riding quality than setts or cobbles at the low speeds recommended for this type of road (50 to 60 km/h) and retains the advantage of being simple and cheap to lay, immediately available for use, and resistant to impact and oil spillage. Blocks also create the minimum amount of trouble when it becomes necessary to attend to buried services; they can be lifted and replaced easily with no consequential disturbance of the appearance of the surface. A specification for precast concrete paving blocks has been produced jointly by the Cement and Concrete Association, the County Surveyors' Society and Interpave, and makes recommendations on materials to be used, dimensions and tolerances and suitable tests. A British Standard Specification is also in preparation.

Construction of a block pavement does not involve the use of specialised paving plant. It consists of a sub-base, laying course and the surface course of blocks. It is also necessary to provide an edge restraint, usually standard precast concrete kerbs, to prevent the blocks from moving sideways (see Fig. 20.3).

The sub-base material is similar to that used for any other road construction, i.e. broken stone or concrete, and for minor residential roads requires a thickness as set out in Table 20.2.

Table 20.2

Type of sub-grade	Thickness of sub-base (mm)
Heavy clay	400 (550)
Silt	400 (550)
Silty clay	190 (300)
Sandy clay	140 (230)
Well-graded sand or sandy gravel	80 (80)

The figures in brackets indicate the depth needed if the water table is less than 600 mm below formation level.

Alternatively, a 75 mm slab of concrete can be laid directly on the sub-grade. This alternative is to be preferred if the initial purpose of the pavement is a haul road for construction traffic.

The laying course consists of clean sharp sand spread and levelled to follow the line of the final surface and compacted to a minimum thickness of 50 mm.

The surface course is the paving blocks. For residential roads,

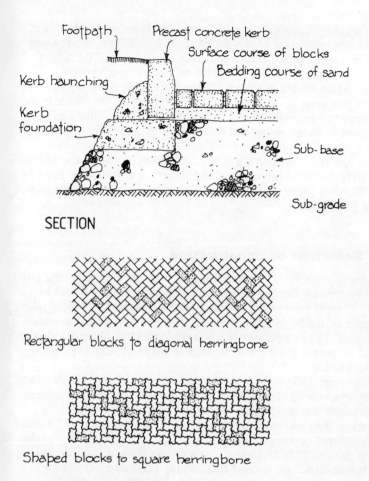

Footpath

Precast concrete kerb

Surface course of blocks

Kerb haunching

Bedding course of sand

Kerb foundation

Sub-base

Sub-grade

SECTION

Rectangular blocks to diagonal herringbone

Shaped blocks to square herringbone

LAYING PATTERNS

Fig. 20.3 Block paving

drives and anywhere where light traffic is anticipated, the blocks should be not less than 60 mm thick. An important feature of the construction is that the blocks must interlock. There are ranges of shaped blocks which interlock with each other and rectangular blocks which are laid to a herringbone pattern (see Fig. 20.3).

Laying commences at the entrance to the paved area. This allows the pavior to stand on the previously laid blocks to work and deliveries of blocks to be made over the completed surface. The blocks are placed dry, by hand, firmly against each other to the specified pattern on the sand bed until a sufficiently large area, including any cut blocks at the edges, has been completed, to allow the next stage of consolidation to start. This is carried out with a plate vibrator consisting of a steel plate, 0.2 to 0.3 m² in area, on which is mounted a petrol engine-driven mechanism. The action of the drive mechanism causes the plate to vibrate in a vertical direction. After an initial vibration, the pavement is lightly covered with sand followed by two or three more passes of the vibrator to agitate the sand into the joints. When the joints are filled, any surplus sand is swept away and the road is immediately ready for traffic.

Although the construction is dry, the filling between the blocks very quickly becomes impervious and therefore the surface must be laid to falls and properly drained in the same way as tarmacadam or concrete pavements.

20.6 Selection of construction

The technology of road construction has now developed to the point where any type of flexible pavement (excluding unbound and block paving), or rigid pavement, can be laid over any sub-grade to take any degree of traffic intensity. Block paving is only suitable where low speeds (but not necessarily low axle loads) are anticipated, and unbound or gravel paving is restricted to areas of very low speed and light loading.

Since either tarmacadam or concrete is suitable for estate roads, the final selection is based on economic grounds and which system of construction fits most readily into the contractor's work organisation. For instance, the extent of the pavement may be small and the general building contractor considers his firm to be capable of laying the road. In this case a concrete pavement is more likely to be chosen (since he is used to handling the material) than a tarmacadam road, the construction of which calls for specialised knowledge and equipment.

20.7 Joints in rigid pavements

Concrete, in common with most building materials, expands and contracts with variations in the temperature. To ensure that the pavement stays flat, this movement must be accommodated with expansion and contraction joints arranged transversely across the road. Since there is an initial shrinkage in concrete as it cures, contraction joints will also take care of a certain amount of subsequent expansion simply by the closure of the joint. For this reason, only every third joint along a road needs to be an expansion joint, the rest being contraction joints.

Strictly speaking, the contraction joint is not a joint at all – it is a controlled crack. The line of the crack is pre-determined by a crack inducer or triangular timber fillet laid on the sub-base. Water is prevented from entering the crack by a bituminous seal run in a groove above the line of the crack inducer (see Fig. 20.4). To keep the surfaces in line each side of the joint, steel bars are cast in, as shown, with one end of each bar treated to prevent it from bonding to the concrete, thus allowing lateral movement to take place, but restricting vertical movement.

Expansion joints are properly constructed junctions between adjacent slabs and incorporate a compressible joint filler strip below a bituminous seal (see Fig. 20.4). A dowel bar, de-bonded for half its length, is cast in, in a manner similar to a contraction joint, but it is fitted with an end cap half filled with a compressible material to allow the concrete to slide up the bar.

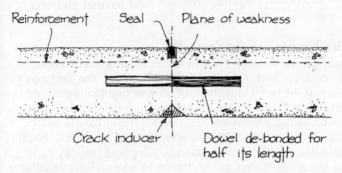

CONTRACTION JOINT

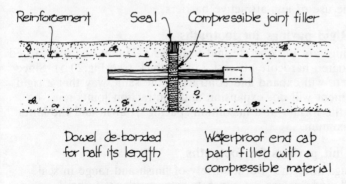

EXPANSION JOINT

Fig. 20.4 Joints in rigid pavements

20.8 Footpaths

Footpaths may be run alongside roads, but in many current residential developments there are segregated pedestrian areas and spine routes (see Fig. 18.5). Whatever the plan arrangements, the paved surface must be firm, sufficiently smooth to allow the easy passage of small wheeled trolleys, prams etc., and so arranged that surface water is not allowed to form ponds. For the purpose of drainage, paths are usually laid with a cross-fall of 1 in 60 towards the gutter or away from buildings. It is also desirable that the selected paving should present an attractive appearance, and recent research has suggested that pavings with an identifiable texture should be used as a guide for blind pedestrians to indicate significant points such as road crossings.

The materials available for these purposes are many and varied, and can be mixed to obtain the desired standard of appearance. They can be divided into flexible pavings, rigid pavings and unit pavings.

20.8.1 Flexible pavings for footpaths

Tarmacadam, similar to road pavements, is used in many areas, usually in two layers laid directly on the sub-grade. The first layer or base course is 40 to 50 mm thick of 25 mm nominal aggregate tarmacadam. Over this is laid a 25 mm wearing course of 10 mm tarmacadam or cold asphalt (see Fig. 20.5). Gravel paving can be used, similar in most respects to the specification for drives, but it tends to make the wheeling of prams and shopping trolleys more difficult and the small stones get scattered by use and the attentions of children. Larger stones, in form of cobbles or flints, usually set in a cement mortar, are used in paved areas but generally to discourage or direct pedestrians' movement at hazardous points without the use of less attractive barriers.

20.8.2 Rigid pavings for footpaths

Rigid pavings are of concrete laid insitu as a 75 mm bed over a bed or consolidated hardcore at least 75 mm thick (see Fig.20.5). Since this concrete will expand and contract in the same way that a rigid road pavement will do, expansion and contraction joints must be provided, the former at a maximum spacing of 27 m and the latter at 3 m maximum.

20.8.3 Unit pavings for footpaths

Unit pavings offer the widest variety of finish and range in scale from small stone or precast concrete setts or blocks to brick pavings and to stone or precast concrete paving flags up to 900 × 600 mm in size (see Fig. 20.5).

Granite setts are square or rectangular blocks of the stone, and

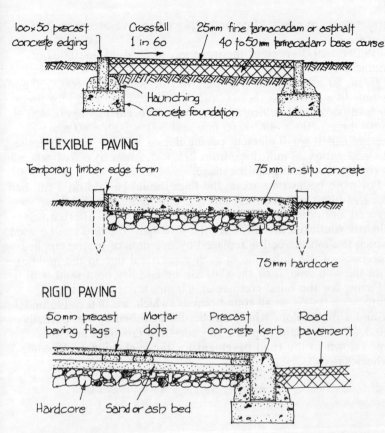

100×50 precast concrete edging
Crossfall 1 in 60
25mm fine tarmacadam or asphalt
40 to 50 mm tarmacadam base course
Haunching
Concrete foundation

FLEXIBLE PAVING

Temporary timber edge form
75 mm in-situ concrete
75 mm hardcore

RIGID PAVING

50 mm precast paving flags
Mortar dots
Precast concrete kerb
Road pavement
Hardcore
Sand or ash bed

UNIT PAVING

Fig. 20.5 Footpath details

for a footpath are laid dry on a 50 mm sand bed with a 10 mm joint left between the setts. After consolidating the paving by rolling or ramming, the joints are filled with cement and sand mortar. They produce a very hard-wearing surface and are sometimes used in conjunction with road pavements as an upstanding kerb or laid flush with the wearing surface to contain the pavement without restricting traffic flow.

Precast concrete block pavings are a lighter version of the blocks already described for roads. The blocks are 60 mm thick and of varying shapes and sizes. They are, like road paving, laid dry on to a 50 mm bed of sand, spread on the sub-grade, sand is brushed into the joints and the paving consolidated by rolling and ramming.

Paving flags are frequently used for footpaths alongside road pavements but also for independant paths. They require no edge restraint and are easily lifted and replaced to give access to buried services.

BS 368 : 1971 specifies the materials to be used, methods of manufacture and physical properties of paving flags. Two thicknesses are given, 50 and 63 mm, and four sizes of 600 × 450, 600, 750 and 900 mm. In addition to these sizes, paving flags are made for areas where an occasional vehicular over-run may occur in sizes of 450 × 450 × 70 mm, 400 × 400 × 65 mm and 300 × 300 × 60 mm – these are called small element paving flags – and for light domestic paths and patios 38 mm thick from 225 × 225 mm to 600 × 600 mm and circular and hexagonal in shape.

For main pedestrian areas, the flags should be laid on a full bed of 3 : 1 sand/cement semi-dry mortar. The joints should be staggered and pointed with 4.5 : 1 sand/cement mix. Alternatively for lightly trafficked areas, either the 'five spot' method can be used, in which the mortar bed is replaced by five dots of mortar applied to the back of the slab (one in each corner and one in the middle) before the slab is laid, or the slab can be laid dry on a sand bed.

Paving for the blind consists of 400 mm square flags with a texture of thirty-six small round bumps, which are detectable under foot and a pink colour which can be distinguished by the partially-sighted. They are laid opposite to pedestrian crossings and are ramped down to the road pavement for the convenience of people in wheelchairs.

Chapter 21

Planning for vehicles and pedestrians in estates

21.1 The main objectives

Estate and urban roads serve many more functions than other types
of main road. A motorway, for instance, is solely dedicated to fast,
efficient transport between main centres. Any other use is banned in
the interests of efficiency and safety. Estate roads must satisfy the
varying patterns of movement of vehicles at lower speed and
frequent stops; other forms of transport; pedestrians and pedestrian-
controlled trolleys and prams. They must do so in a safe, convenient
manner which avoids creating a nuisance for the estate residents and
contributes to the appearance of the environment. In any conflict of
interests, the needs, safety and convenience of pedestrians should be
given priority over the demands of vehicles.

To implement these policies the following main objectives should be
observed:

(a) The layout must achieve an economic balance between the scale
of provision and the minimum commensurate with safety,
convenience and cost-in-use.
(b) Non-access traffic should be discouraged, but where it is
unavoidable any danger or nuisance should be minimised.
(c) The amount of vehicle flow and its speed should be kept to low
levels in the vicinity of houses.
(d) The routes for vehicular movement must be safe.
(e) On-street parking should be discouraged to minimise danger to
pedestrians and inconvenience to emergency and other services.

(f) All routes should be clearly and adequately signed.

(g) Safe and convenient pedestrian routes should be provided between all the houses and facilities such as bus stops, shops, playing fields and other community provisions.

(h) Public services must be accommodated by a means which is adequate for the residents' needs and convenient for maintenance purposes.

(i) The overall appearance of the development must meet satisfactory environmental standards.

21.2 Economics of road layouts

The provision of the roads and footpaths in a residential development can account for as much as 10 per cent of the contract costs. This aspect of the development also leads to a substantial, recurring, annual cost-in-use in the form of maintenance and lighting expenses.

Skilful planning in the first instance can keep to a minimum the paved areas needed to provide a service to each house. The grouping of houses round a small service mews and the careful selection of pavings and kerbing can reduce the overall cost of the roads, thus releasing more money for the houses.

Economies in the construction must not be made at the expense of the long-term maintenance and repair costs of the road. All construction methods adopted must be adequate for the projected traffic load over the design period – usually 40 years – but in any given set of circumstances there is usually one choice which is more economic in materials or labour than any other.

A difficult matter to resolve is the balance to be struck between the probability of damage occurring due to the worst possible combination of the worst possible behaviour of drivers and pedestrians and the capital cost involved in combating these worst excesses.

An examination must be made of the savings that can be effected by limiting standards to meet normal behaviour patterns and the cost of repairs needed to restore the road and footpath areas, should damage occur.

21.3 Non-access traffic

Any estate design should, for the convenience of the residents, be so arranged that it is impossible or inconvenient to use the residential roads as a short cut between two distributor roads. Such use has the effect of increasing both the volume and weight of traffic and its speed (the drivers are not intending to stop and, indeed, are seeking to reduce their journey time) and these are accompanied by the nuisances of more fumes, noise and vibration, plus the additional hazards created.

It is far better that the deterrent should be self-enforcing by the design of the road system than that it should have to rely on the use of signs. Figure 21.1 shows typical road layouts in which the through traffic is encouraged to stay on the distributor routes.

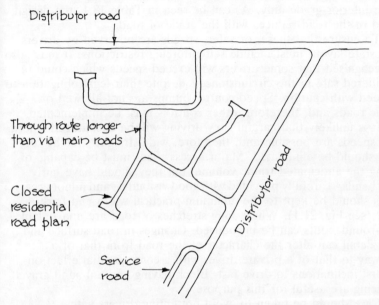

Fig. 21.1 Planning to deter non-access traffic

While the ideal target for which to aim is to have all houses served from a quiet residential road and remote from busy through routes, the economics of land use and other reasons can create situations where houses have to be alongside distributor roads. In such cases, small service roads can be constructed to avoid direct access between the houses and the main road.

21.4 Controlling traffic flows

The principal reason for controlling the amount and speed of traffic flow on estates in the vicinity of the houses is to avoid, or at least reduce, accidents, particularly to children. The proportion of accidents involving children compared to those involving adults is highest on roads solely serving houses and half of all accidents to children under five occur within 100 m of their homes.

The volume of traffic is related to the number of houses directly served by the road, and the speed must be assumed to be the maximum possible or safe in the conditions provided. In any estate development the number of vehicles travelling along the roads at the entrance will be much greater than those found in the more remote parts of the layout. For this reason a hierarchical system of roads has been developed, as shown in section 18.5.1, in which one superior grade of road leads to either another of equal grade or one of an inferior grade only. As can be seen in Table 18.1, the design speed of the road reduces with the grade of road.

To ensure that these, or safer, speeds are observed by drivers, the estate layout must include self-enforcing restrictions. It must also be recognised that some drivers will exceed speeds which could be considered safe in the circumstances, despite their legal obligation to proceed with caution. Speed control bumps are not allowed on public roads and, therefore, other controls must be implemented.

It is unlikely that careful slow driving will be encouraged where high speeds are possible and, therefore, wide straight stretches of road should be eliminated. Major access roads must be capable of taking the anticipated traffic volume, but they should have fairly tight bends – carefully provided with good visibility – and minor access roads should be kept to the minimum practical width and fairly short (see Fig. 21.1). Where long stretches of road are unavoidable, mini-roundabouts can be introduced. Changes in road surface and kerb detail can alter the character of the road from that of a highway to that of a private drive, with a consequential effect on drivers' inclinations to drive fast. Block paving and unbound gravel surfacing are useful for this purpose.

Care should be taken to avoid presenting drivers with unexpected conditions or situations they cannot anticipate, either of which could constitute a safety hazard, nor should the restrictions designed to slow down private cars make it so inconvenient for larger vehicles that the quality of service to the houses is degraded.

21.5 Safety of vehicular routes

For routes to be safe, they must accommodate the physical size of the largest vehicle likely to use them. Due regard must be paid to the vehicle's width, the tolerances required each side of it, the space it takes up when turning a corner, safe visibility and, where necessary, what provision must be made for it to turn round.

The vehicles which use estate roads are private cars, delivery vans and lorries, refuse collection vehicles, removal vans and fire appliances. The largest of these is the removal van. Very large lorries are generally confined to main distributor roads and, although there are some fire appliances which need a lot of space, they are only needed for fires in high-rise buildings and, therefore, would not appear on minor estate roads.

A typical removal van is approximately 9.5 m long, 2.5 m wide and 4.0 m high, and it describes a turning circle of about 20 m between kerbs or 2 m between walls. The turning circle between kerbs is the line followed by the front outer wheel and that between walls is the line followed by the front corner of the vehicle body. Figure 21.2 shows typical allowances to be made for a removal van, with that of an average car for comparison.

With the low speeds to be encouraged in an estate layout, an overall difference of 0.5 m between vehicle width or widths and the carriageway is considered adequate. Where speeds are greater, as on trunk roads and motorways, this tolerance must be increased. The absolute minimum road width is, therefore, 3.0 m where a one-way or single-track system is used (single-track roads are intended for two-way traffic but consist of single-width tracks with passing bays at intervals). At 4.1 m, cars could pass each other and a removal van could pass a cyclist but not a car. A width of 4.8 m is required to allow a car and a removal van to pass and 5.5 m to allow completely free movement of all vehicles. This last figure is the width shown in Table 18.1.

None of these widths allows for casual on-street parking which, although it is to be discouraged, must be anticipated in roads giving direct access to houses (see section 21.6 for parking requirements).

As can be seen in Fig. 21.3, the space occupied by a vehicle when turning is greater than when it is travelling in a straight line. In recognition of this, bends in some roads should be widened to allow vehicles to pass each other.

It is also necessary to ensure that the driver of any vehicle negotiating the bend can see clearly any approaching traffic or other hazard. To allow for this, no obstructions to sight should be allowed between the inside line of the bend and a forward visibility curve. Figure 21.3 shows a forward visibility curve to a 90° corner of 25 m radius and is constructed as follows:

(a) Draw the path of the vehicle 1.5 m from the inner kerb.
(b) Read off the anticipated mean speed for carriageway radius from graph A (in this example 30 km/h).
(c) Read off the stopping distance for mean speed from graph B (in this example 32 m) and round it up to a multiple of 3 m (i.e. 33 m).
(d) Plot this distance along the vehicle path from tangent point A and divide into numbered 3 m intervals.
(e) Set out the same number of numbered 3 m intervals round the curving vehicle path and again along the straight path beyond tangent point B.
(f) Connect increments of the same number to give an interlacing pattern of lines covering the area to be kept clear.

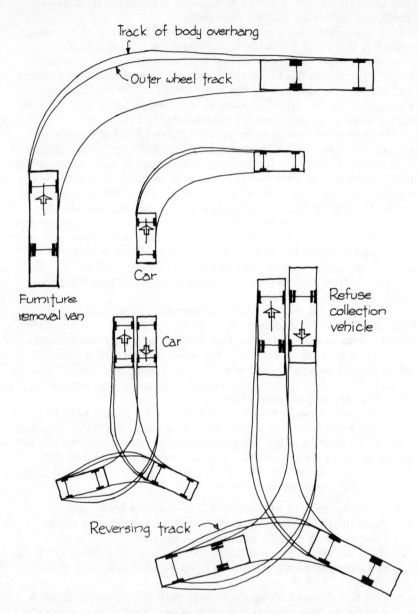

Track of body overhang

Outer wheel track

Car

Furniture
removal van

Refuse
collection
vehicle

Car

Reversing track

Based on DoE & DoT Design Bulletin 32

Fig. 21.2 Comparative turning circles

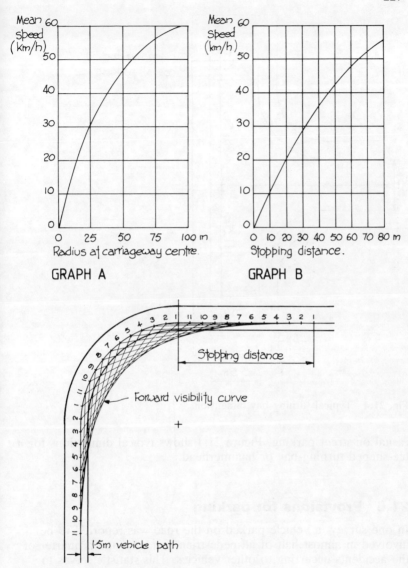

Fig. 21.3 Forward visibility curve

Visibility at junctions is catered for by the provision of sight splays which vary depending on the types of connecting roads (see Fig. 18.3).

Turning bays must be provided at the end of culs-de-sac and must be large enough to accommodate a vehicle such as a refuse collection freighter (see Fig. 21.1) and, where it is likely to occur,

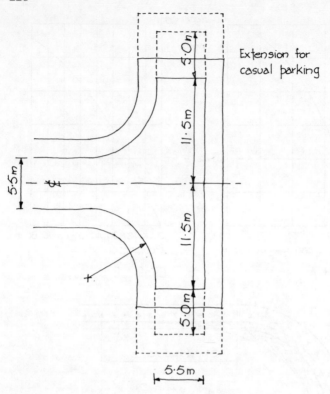

Extension for casual parking

Fig. 21.4 Typical turning bay dimensions

casual on-street parking. Figure 21.4 shows typical dimensions for a tee-shaped turning bay or hammerhead.

21.6 Provisions for parking

In one survey, a vehicle parked on the road was reported to be involved in almost half of all pedestrian accidents and a quarter of the accidents occurring to other vehicles. This statistic serves to emphasise the importance of inducing drivers to leave the carriageway to park. To do this, the parking areas must suit the needs of residents, visitors and service and maintenance vehicles for both short-term and long-stay parking. In addition, the routes between the parking spaces and the entrances to the dwellings must be shorter and more convenient than would be the case if parking were on the carriageway.

Parking can be either long-term or short-term and can involve residents' cars, visitors' cars and service vehicles–thus there are six different parking needs to be met. In developments of detached or semi-detached bungalows and houses, residents' and visitors' parking and garaging is usually accommodated within the curtilage of the site. Where terraced housing and flats are planned, the parking provisions are often grouped into hardstandings which must be as near to the entrance as possible and certainly nearer than the road.

Long-stay service vehicles such as removal vans or builders' lorries do not occur with a regularity which justifies a specially provided parking area, but it must be recognised that the need will arise from time to time and it should be possible to place such a vehicle at a convenient distance (10 to 30 m) from the house without total obstruction of the road or the creation of a serious hazard. Short-stay service vehicles such as refuse-collection freighters and milk floats present a constantly changing obstruction and should be able to stop within a distance of 25 m to the delivery or refuse collection point (9 m if the refuse is in a large communal container).

Fire appliances must be able to get within 45 m of one of the entrances to one- or two-storey houses and 36 m of three- and four-storey blocks. Higher blocks of flats will require special consultation with the fire officer.

The minimum size of a parking space for a private car is 5 × 2.5 m. Where a structure adjoins the space, additional room needs to be allowed for opening the car door and so the width should be increased to 3 m. An additional allowance of 1 m in the length must be made if the parking space is in front of a garage to allow the doors to be opened, thus increasing the length of the space to 6 m. To allow visibility when emerging on to the road, sight splays of 2.1 m must be arranged where a building or a wall above 0.6 m adjoins the entrance. All these dimensions are shown in Fig. 21.5.

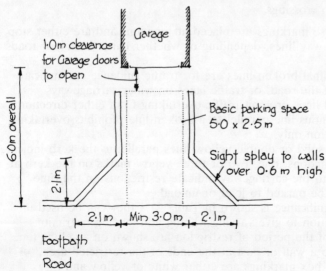

Fig. 21.5 Parking requirements

The minimum internal size for a garage is 4.9 × 2.4 m. Where they are grouped into garage courts, there should be a minimum of 7.6 m between opposing doors and the width of access should be between 2.5 and 6 m according to the number of vehicles garaged.

Where access to the parking space or garage is from a Type 2 road (see Table 18.1), allowance must be made within the site of the house for the car to be turned round so that it can both enter and leave in a forward direction.

21.7 Road signs and markings

Drivers are given directions and information by road markings and by various signs fixed to posts or walls. The form, size and position of these marks and signs are closely controlled by law. The amount of signing required on estate roads tends to be limited. The following is a selection of the most common — further information can be found in *The Traffic Signs Regulations and General Directions 1975 (Statutory Instrument 1975 No. 1536)* by the Department of Transport.

21.7.1 Road markings

There are six types of marking carried out:

1. Transverse white lines.
2. Longitudinal white lines.
3. Yellow markings for waiting and loading restrictions.
4. Worded and box markings.
5. Junction markings.
6. Pedestrian crossings.

Transverse markings are placed at junctions and are either stop lines or give way lines, depending or whether major or minor roads are involved.

Longitudinal broken lines are for traffic guidance and indicate the centre of the road, or traffic lanes on wider carriageways. Double solid lines prohibit traffic overtaking from either direction and a continuous line parallel to a broken line prohibits overtaking in one direction only.

Broken, solid or double yellow lines parallel to the kerb indicate parking restrictions; one, two or three yellow marks on the kerb at right angles to the carriageway indicate restrictions on the time vehicles can be parked to load or unload.

Their significance is shown in Table 21.1 they can be used in any combination to give varying waiting and loading restrictions.

Details of the period of restriction are shown on a yellow time plate fixed to a wall or post at the ends of the restricted area. Worded and box markings are either white or yellow and give

Table 21.1

Mark on road	Mark on kerb	Restriction period
Single broken line	Single mark	Peak hours only
Single continuous line	Two marks	7.00 am to 7.00 pm
Double continuous line	Three marks	7.00 am to 7.00 pm plus any period between 7.00 pm and 7.00 am

instructions (STOP), warning (SLOW) or indicate areas of carriageway used for a special purpose (BUS STOP).

Junction markings are white and comprise a stop line, lane markings and direction arrows.

Pedestrian crossings on estate roads are usually uncontrolled, i.e. they do not have traffic signals and are marked with the familiar alternate black and white stripes of the zebra crossing. Zig-zag lines are drawn for a distance of 18.75 m on each side of the crossing to indicate a no-waiting area. Controlled or pelican crossings do not have black and white stripes, but are marked by two rows of studs plus a double row across the approach lane, 15 m from the crossing.

The most commonly used material for marking estate roads is superimposed thermoplastic laid hot at a temperature of 130°C. Its durability is enough to last the interval between road surface redressings with the intensity of traffic found on estates and is less expensive than the more durable materials used for heavily trafficked roads.

A cheaper and less durable material is road paint. It can be applied by machine using a spray technique but is usually hand painted because it is this simplicity of application which is its main advantage. Marking of block pavements is achieved with the use of coloured blocks set as part of the surface to give required lines etc.

21.7.2 Road signs

There are three main classes of road signs, reflecting their differing function: regulatory signs, warning signs and information signs.

All regulatory signs are circular, with the exception of the STOP sign which is enclosed in an octagon and the GIVE WAY sign which is an inverted triangle. They give notice of requirements (KEEP LEFT), prohibitions (NO ENTRY) or restrictions (NO WAITING).

Warning signs use a symbol enclosed within a triangle and may have a rectangular plate below to give additional information, e.g. two children enclosed within a triangle give warning of a school further along the road. This message may be reinforced by a plate below the triangle with the word SCHOOL.

Information signs, generally, comprise directional signs, which assist drivers to find their route, and street name-plates. Since a well-planned estate does not have any routes through it, the street name is the only type of this class of sign found on these roads.

The siting of signs is determined by its purpose, the speed of the traffic and the road feature to which it applies. Most signs are mounted on posts at a minimum distance of 0.5 m from the edge of the carriageway, with the lower edge between 1.0 and 1.5 m above the carriageway and at 95° to the line of travel of the traffic. Direction signs, waiting restriction signs and street name-plates are mounted parallel to the carriageway on posts or a convenient wall. All posts should be set in the ground a distance equal to one-third of the height above ground and surrounded in concrete.

21.8 Appearance

The overall appearance and the way the estate is looked after are two of the most important factors which determine satisfaction among the people living in the houses. Both must be considered when planning, the appearance being studied to ensure a pleasant welcoming character, suitably small in scale, and the maintenance must be kept to the minimum by a careful selection of materials and planting arranged in such a way that the inevitable wear and weathering is either unobtrusive or an acceptable part of the maturing of the landscape and environment.

21.9 Pedestrian routes

People like to take the shortest, most practicable route when walking between two points; if the footpath does not follow this route, they will create their own short cut (see Fig. 21.6).

The footpath must also be safe. The preference is to segregate vehicular and pedestrian routes but where both the road and the footpath give access to the houses, they inevitably come together. In this case, raised kerbs or other barriers must be provided to discourage or prevent vehicles mounting the footway. Grass verges can also provide a safety margin between the path and the road.

For convenience, the footways should be wide enough for two prams to pass. A minimum of 1.7 m is required for this purpose. This should be increased to 1.8 m for ease of movement over long distances and 2 m if the path is beside a road to avoid any need for passers-by to step on to the carriageway. Any ramps should not be steeper than 1 in 12, preferably less, for the benefit of wheelchair users and, as described in section 20.8.1, special textured pavings are an assistance to the blind or partially sighted.

When the purpose of the pedestrian route is to gain access to a particular service point – a shopping centre, bus stop, school,

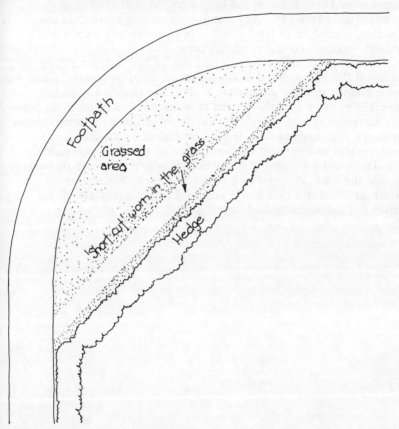

Fig. 21.6 Short cuts in footpath layout

recreation area etc. – a direct, segregated spine route can be
provided. This can be completely safe and pleasant to use. Figure
18.5 illustrates a typical layout showing a pedestrian spine route
leading straight from the local distributor road to the central
shopping area.

21.10 Accommodation of public services

The routes followed by drainage, gas, water, electricity and
telephone services is, to a large extent, dictated by the
configurations of the road layout. Whatever the arrangement, the
service runs must be adequate for the demand anticipated and cover
the shortest distance possible. However, the most economical route

may not always be the best for the purposes of maintenance and hence would cost more in the long term, as well as causing annoyance to residents. Any installation serving more than one dwelling should be in land which is both commonly owned and publicly maintained and is readily accessible.

Where possible, services should not be laid below carriageways, to avoid the expense and inconvenience caused by gaining access for maintenance in these circumstance. It is not usually completely possible to avoid branch services crossing the line of the road below the carriageway but where appropriate these can be accommodated in ducts, thus making their initial installation and subsequent replacement independent of disturbance of the road pavement.

The usual position for all services is below the verge or footway beside the road, grouped in a common trench. For this purpose a width of about 1.5 to 1.8 m is required to accommodate all the pipes and cables to be laid (see Fig. 22.2).

Chapter 22

External services

22.1 Services to be provided

As mentioned in the last chapter, the designing of an estate must also take into account the best route for the services. This may not always be the shortest route for reasons of access for maintenance, but it is usually followed by all the services except main drainage.

The routes to be used by main drainage runs are inflexible both in plan, since they must be a series of straight lines, and in section, because there must always be a fall in the length of each pipe. Because of this inflexibility, public and private sewers may not be found running along with other services.

The other services normally provided are a water main, a low-pressure gas main, a low-voltage electrical cable and a telephone cable. In addition to these, the area may be served by a television relay system or a district heating scheme, and it may be crossed by high-pressure gas mains or high-voltage electricity mains cables.

22.2 Service points

Wherever possible, service routes should be in land maintained by the local council at public expense, but placing pipes and cables below road pavements should be avoided because of the inconvenience and higher reinstatement costs which arise when the carriageway has to be dug up for service maintenance or repair. It therefore follows that most service pipes and cables are laid below footpaths, verges and public open spaces.

In the case of drainage systems, pipework laid below public land, serving two or more houses and maintained by the local authority, is termed a public sewer. It is a duty of local authorities to provide public sewers and sewage disposal works and they can require any building within 30 m of a public sewer to be connected to it.

Some estate layouts show several houses connecting to one common drain which runs through the gardens of each property before it connects to the public sewer. This is a private sewer (see Fig. 22.1). All property owners must grant a right of drainage to the upstream owners and everybody connected to the private sewer is collectively responsible for its maintenance. Legal agreements to this effect must be included in the deeds of each property, but even then arguments can develop when repair work has to be carried out to correct a fault, affecting only one or two houses at the head of the sewer, but which must be paid for by all property owners.

In some estates there is a dual system of sewers, one for soil disposal (the water from WCs, baths, basins and sinks) and one for surface water (from rainwater pipes and road gullies). These may be routed together but they may follow quite independent lines to separate points of connection or disposal.

With not having to follow straight lines and even gradients, the other services can be laid below the footpath or verge and easily remain alongside the line of the road no matter what its route. If they are to run parallel to each other, it is sensible to co-ordinate the installation work so that it can all take place in a common trench. To allow for subsequent access to an individual pipe or cable, they should be arranged in the manner shown in Fig. 22.2. The best position for this common trench is below a grass verge since access causes the least obstruction and reinstatement is the cheapest. Where a grass verge does not exist, the installation should be below the footpath.

Main service runs are laid down one side of a road only but, as houses are built along both sides, it is necessary for branch connections to cross the carriageway. These usually run in p.v.c. or pitchfibre pipe ducts laid in the sub-grade before the road is constructed.

22.3 Drainage systems

As mentioned above, in some estate layouts there may be a dual system of drainage, one sewer for soil water and one for surface water; in others, both types of effluent may be run in the same; combined, sewer.

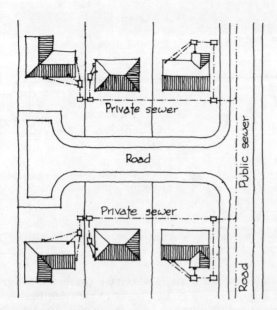

PRIVATE SEWER LAYOUT

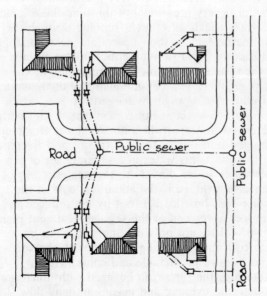

PUBLIC SEWER LAYOUT

Fig. 22.1 Public and private sewers

238

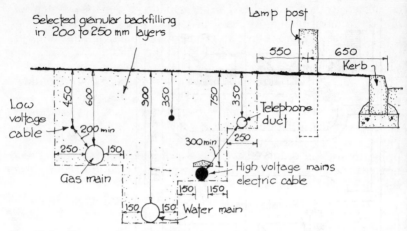

Fig. 22.2 Common trench details

The problem facing local authorities who have combined sewer
layout is the sudden large flow which can occur in wet weather
conditions. There are two difficulties to be faced in this situation.
Firstly, the sewer pipes must be large enough to take this sudden
rush of water but, as this only occurs at its very maximum only
every five years or so, for a very great deal of the time these large,
expensive pipes have just a trickle of effluent running through them
which tends to lead to inefficiency in their operation (near-horizontal
pipes operate most efficiently when running either full or exactly
half full). Secondly, when the sudden flow of storm water reaches
the sewage treatment plant there is a considerable risk of the system
being overloaded and untreated sewage discharged.

Since the storm or surface water is not contaminated with bodily
wastes and other deleterious matter, it does not need any treatment
before it can be discharged into our system of streams and rivers. It
is sensible, therefore, to collect this water in a separate set of pipes
and to run this surface water sewer directly to the nearest
watercourse. The main disadvantage is the additional cost of the
second system of sewer pipes, but this is offset by the reduced
capacity needed in the soil sewer and at the sewage treatment plant.

A further advantage is that, not only can the soil sewer be
smaller, thereby saving cost and increasing its efficiency, the soil
discharge volume is much more predictable and constant than
surface water and, therefore, the pipes can be sized with much more
precision to accommodate the average and maximum daily flow. In
the long term there are also substantial cost-in-use benefits to be
gained from the reduced operating costs of the smaller sewage
treatment plant.

Clearly, with a separate system of sewers, the buildings

connected to them must have a correspondingly separate system of drains. In most building designs, the soil drainage points are grouped together to achieve an economy in plumbing and drainage but, since it rains equally on all parts of the building and site, the surface water drainage points cannot be so conveniently grouped. This dispersal of surface water drainage points makes the building drainage system extensive and costly and, in some cases, not physically possible. To alleviate this problem, some local authorities have a partially separate system in which there are two sets of sewer pipes, one for surface water only and one which is intended primarily for soil drainage but into which may be discharged a limited amount of surface water from points which are remote from the surface water drains.

Where a satisfactorily permeable subsoil exists, some authorities provide just a soil sewer and require the surface water to be discharged into soakaway pits or, if convenient, to a ditch. Soakaway pits consist of either holes dug in the ground filled with clean broken stone or brick, or specially made permeable tanks set in the ground which act as catch pits at the end of the drains to collect the rainwater and allow it to seep away into the ground.

Figure 22.3 shows diagrammatic layouts for the systems described above.

22.4 Drainage materials

There are four materials commonly used for drain pipes and fittings: vitrified clay, concrete, unplasticised p.v.c. and pitch fibre.

22.4.1 Vitrified clay pipes

These, originally known as salt-glazed stoneware pipes, are made from a mixture of mineral materials such as sand, and a clay binder with a high kaolin content. This material has been used for drainage goods for a very long time, but in recent years the system has undergone considerable improvement in the methods of jointing. Formerly, pipes were connected by a rigid joint made with cement and sand; now flexible joints are used which permit slight movements to take place in the drain line without any leaks developing. These movements are inevitable and are due to earth movement and thermal changes.

The pipes are made in either the traditional manner with a socket on one end which receives the plain or spigot end of the next pipe, or a simple straight barrel plain-ended pipe.

The jointing methods used employ either plastic bedding rings and rubber sealing rings within the spigot and socket joint, or a polypropylene sleeve pipe which fits over two adjacent plain-ended pipes (see Fig. 22.4).

240

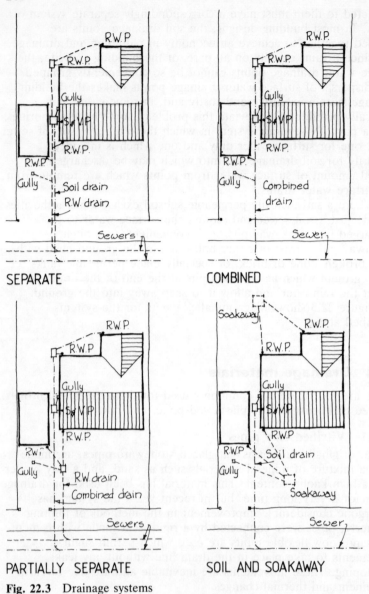

Fig. 22.3 Drainage systems

They are usually stocked by merchants in sizes of 100 and 150 mm diameter but are available from 75 to 750 mm diameter and 600 to 1500 mm long.

22.4.2 Concrete pipes
Concrete pipes are often made by spraying the mix on to a rotating

241

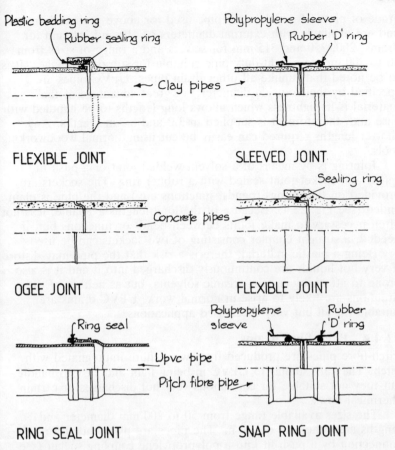

Fig. 22.4 Drain pipe joints

former and are particularly suitable for very large diameter drains (up to 1800 mm). They could be susceptible to attack by chemicals in the ground or in the drainage effluent but are treated with protective wrappings and linings or made with sulphate-resistant cement.

Jointing is by means of an ogee shaping at the ends of the pipe or by spigot and socket. Ogee joints are suitable for surface water drains only. The spigot and socket joints can be made as a rigid connection with cement and sand but a better practice is to use a rubber sealing ring which is compressed in the joint as the pipes are forced together.

22.4.3 UPVC pipes
Unplasticised polyvinyl chloride pipes have greatly simplified the job of drainlaying. They are coloured golden brown (to distinguish the

grade of plastic from the grey pipe used for above ground drainage) and are available with external diameters of 110 and 160 mm for drains, 200, 250 and 315 mm for sewers and a range of sizes from 50 to 200 mm in a light-duty pipe suitable for subsoil drainage. (It is to be noted that, unlike all other drain pipes, UPVC pipes are specified by their outside diameter.) A great advantage of the material is its lightness which allows long lengths to be handled with ease and, therefore, it is supplied in 3.0 and 6.0 m lengths. Any shorter lengths required can easily be cut using normal woodworking tools.

Jointing is by means of a solvent-welded joint or a push fit spigot and socket joint sealed with a rubber ring. The sockets are provided on the fittings (bends, junctions etc.) and the pipe is plain ended (see Fig. 22.4). Because of the long lengths available, it is not often necessary to make joints between bends or junctions but, if needed, a straight coupler consisting of two sockets can be used.

Being a plastic material, there is a risk that the pipe may distort if very hot liquids are continuously discharged into it and it is also prone to attack by certain organic solvents, but as neither of these situations are likely to arise in normal work, UPVC drains are suitable for all but very specialised applications.

22.4.4   Pitch-fibre drains

Pitch-fibre pipes are produced from wood fibre impregnated with pitch. They are similar to UPVC in being light and easy to handle, but they not suitable for continuous very hot discharges or certain chermical flows.

The sizes available range from 50 to 200 mm diameter and lengths are either 2.5 or 3.0 m. The pipes are plain-ended for connection by a push fit into a polypropylene coupling socket (see Fig. 22.4). When they were first introduced, the pipes were machined with a taper on each end which matched a corresponding taper inside the coupling; when the coupling was driven on to the pipe, the force caused the mating surfaces to seal and no other jointing compound was necessary. This did not give a flexible joint capable of accepting changes in pipe length due to thermal variation and, therefore, tended in time to develop leaks.

22.5   Manhole construction

Manholes or inspection chambers must be provided wherever there is a junction or change of direction in the drain to allow access for the purposes of clearing any blockages. They must be adequate in size and proof against ground water. The traditional form of construction is to lay a concrete slab as the base of the chamber and, on to this bed, half-round section vitrified clay channels and junctions connected to the main drain and branches. Off this slab

243

are built three or four courses of engineering bricks and the space between the manhole walls and the drain channel filled with cement and sand, or concrete benching. The walls are then continued to within about 200 mm of ground-level. The manhole is capped with a 100 mm concrete slab, in the centre of which is a rectangular hole corresponding to the manhole cover size. A course of bricks is bedded on the top of this slab and levelled up to receive the frame of the manhole cover. This is bedded on the levelling course in cement mortar and the cover is bedded into the frame in grease (see Fig. 22.5).

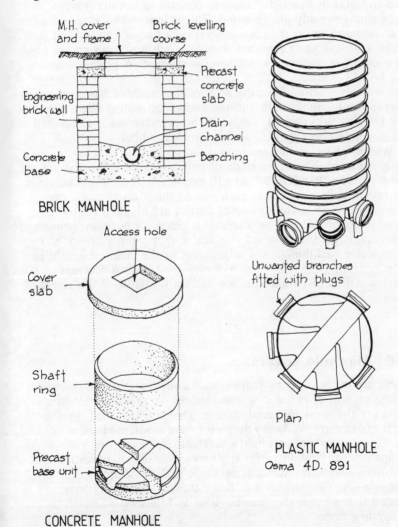

BRICK MANHOLE

CONCRETE MANHOLE

PLASTIC MANHOLE
Osma 4D. 891

Fig. 22.5 Types of manhole chamber

As illustrated, the manhole is suitable for shallow depths only; deeper manholes should be made larger, to allow more ease of movement, and provided with step irons or a built-in ladder for ease of access.

To build a brick manhole is quite an expensive operation and, therefore, alternatives in the form of precast concrete and plastic manholes are now produced. Both are usually circular in plan, although some concrete manholes are rectangular and both are illustrated in Fig. 22.5.

The concrete manhole has a precast base formed to give the main and branch channels, ready to connect to the drain-pipes, with the benching already incorporated. This is lowered by crane into the excavation and set at the correct level on a bed of cement and sand. Precast concrete shaft rings are then lowered in and set on the base and each other until cover slab level is reached. A precast cover slab, holed for access, is then set on top of the shaft. If the water table is likely to rise above the level of the manhole base, the space between the manhole shaft and the excavation wall is filled with at least 150 mm thickness of concrete. The construction is completed by a manhole cover bedded as for a brick chamber.

Where UPVC drains are used, it is the general practice to use a plastic manhole. These are made of polypropylene and arrive on site in one piece, either 220, 570 or 910 mm deep. The base is moulded with two branch connections each side of the main channel, one at 90° and one at 45°. Any unwanted entries are blanked off by the special plugs provided. The manhole is placed in position, connected to the drains via the standard ring seal sockets, and surrounded by a fine granular backfilling. Any adjustment to the manhole height is made by cutting the shaft with a fine-toothed saw. A 150 mm concrete slab is cast round the top and on this is bedded a standard manhole frame and cover.

22.6 Manhole covers

Covers are made in many shapes, sizes and strengths to suit each installation and of either cast iron or steel. In practice, cast-iron covers are the most frequently found. They are classified, by the load they can carry, as: heavy duty – for main roads; medium duty – for areas where only light wheel loads, i.e. a car, are anticipated; and light duty – for areas accessible only to pedestrians. Light-duty covers are available as single seal, illustrated in Fig. 22.5, or double seal for internal use. Some double-seal covers have recessed tops so that they can be filled and finished to match the surrounding floor.

22.7 Sewer connection

It is frequently necessary to connect a new drainage system to an existing sewer. The method by which this is achieved depends on the relative sizes of the sewer and the drain.

Connections of normal domestic size drains to large sewers are made by a saddle piece (see Fig. 22.6). To do this, a small hole is

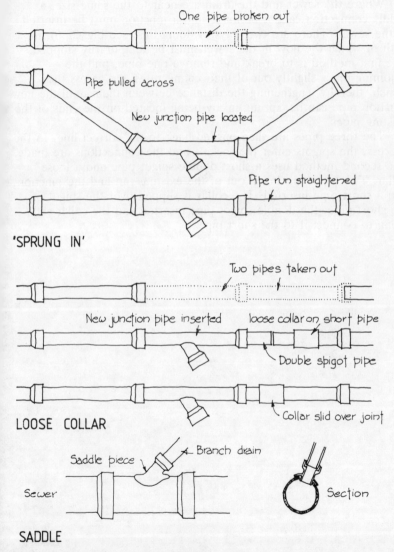

Fig. 22.6 Sewer connection methods

broken through the upper half of the sewer pipe and carefully enlarged until the spigot of the saddle piece can be inserted. The outside of the pipe is thoroughly cleaned and the saddle bedded in place on a 1 : 3 mix of cement mortar. When the mortar has set, the drain is connected, the whole connection surrounded with at least 150 mm of concrete, the excavation carefully backfilled and the surface reinstated.

Where the sewer and the drain are near to the same size, a saddle connection cannot be made and a junction must be inserted in the sewer. There are two ways of doing this (see Fig. 22.6), in both of which the flow in the sewer must be temporarily stopped. The first method is to break and remove one pipe, pull the adjoining pipes slightly out of their connections and across the trench, thereby lengthening the distance between them until the new junction pipe can be 'sprung in' by being located on the ends of the existing pipes.

The three pipes are then pushed back to the correct line. In the process, the spigots enter the sockets and the connections are made. The second method uses a short double spigot pipe and a loose collar. Two pipes are taken out of the existing run and the junction, and a short pipe inserted. The collar is slipped over the spigot of the short pipe, the double spigot pipe inserted and the collar slid along to connect it to the short pipe.

Part D

Organisation of work

Chapter 23

The effect of site features on construction organization

23.1 Site features

Most of the pre-contract planning carried out by the builder is concerned with how to organise the execution of the work on site: how best to use the space or spaces available after allowing room for the building or buildings, the associated scaffolding, service runs and roads. Careful thought must be given to this subject, particularly on congested sites, to avoid having to move stock piles or materials or site accommodation during the course of the contract.

The controlling factors in this respect are, obviously, the proposed layout of buildings, roads and services, In addition, there are a number of site characteristics which influence the final decisions made. Each site presents its own peculiarities, but the features which regularly need consideration are: the location of the site; the means of access both to the site location and from the adjoining roads to the site itself; the size of the site; the size of the contract; and the type of work to be carried out.

These features will affect the site accommodation, as far as its extent, arrangement and type are concerned, and the delivery, storage, control and security of materials and components intended for the construction works. The details of the accommodation which must be provided is dealt with in Chapter 26 and the management of materials is the subject of Chapter 25.

23.2 The location of the site

The problems faced by a builder planning a development on a city-centre site are very different from those posed by a site in a remote rural area.

The city-centre site will almost certainly be congested, with a high percentage covered by buildings, leaving very little room for the accommodation needed for the men working on the site, nor space to store materials and components. It would, however, probably be adequately served by public transport and, therefore, any need for the accommodation of staff cars is likely to be limited in extent or even non-existent.

Rural sites are usually much more open than city-centre sites, the percentage of site coverage much lower and, therefore, site accommodation and storage is much easier to achieve. Public transport is not as good as in built-up areas and, therefore, the men working on the site will have to travel, either in private cars – for which parking must be found – or by specially run company buses.

Deliveries of materials to city-centre sites are subject to the traffic conditions in the roads adjacent and to any imposed control on such things as parking and loading time restrictions, weight limits, height limits etc. In rural areas, the distances to be covered by the delivery vehicles (and the lorries removing rubbish, packaging and items no longer required) are greatly extended and the roads can be narrow and twisting.

Security against theft and vandalism needs to be more stringent in city centres than in the country, where there are fewer people. But, since the quantities of materials which can be stored on city-centre sites is generally much less than on rural sites, the extent of the more stringent security arrangements is correspondingly diminished.

23.3 Means of access

As mentioned above, traffic congestion in cities and narrow lanes in the country can each create their own, special problems. City-centre sites, because of the lack of bulk storage facilities, tend to be developed using prefabricated components delivered to a carefully timed programme allowing components to be lifted off the lorry and placed directly into position. For reasons of building economics, such components need to be as large as possible and their arrival accurately timed to avoid idle crane time. If the adjacent roads are constantly packed with traffic, it may not be feasible to make regular journeys with a large transporter, nor can arrival times be relied upon because of delays occurring en route.

The size of components, and the lorries delivering materials, may also be restricted by the physical impossibility of negotiating the narrow lanes and tight corners encountered on the way to a rural site. Component size is a subject for study by the building designer, but the builder must anticipate any difficulties in this respect. Delivery lorry size is controllable, to an extent, by the builder and he must arrange his ordering and materials storage to take account of any restrictions.

Access from the road to the site must be provided on a permanent basis in all developments. In many, this access is restricted to one point only by either local authority control or by the length of site frontage. Wherever possible, the builder will use the intended permanent point of access as his entrance for construction traffic. If there is only one point allowed, the builder is faced with the choice of allowing the lorries to drive in and then, after unloading, back out on to the road; directing the lorries to back in so that, subsequently, they can drive forwards into the road; or providing a sufficiently large turning area within the site to allow vehicles to drive forwards both into and out of the site. The first choice results in a dangerous situation unless the traffic is controlled while the lorry is reversing; the second also requires a certain amount of traffic control and will inevitably cause congestion; and the third choice requires a free manoeuvring area of a size which frequently is not available on the site.

A better solution to the problem of deliveries but one which may add to security problems is to have two points of connection to the road, one for ingress and one for egress, so that all vehicles can conveniently and easily drive in at one and out of the other. This requires a sufficient length of frontage to one or more roads and the co-operation of the local authority. Another solution is to leave the delivery vehicle on the road and transport its load into the site by other means such as a dumper truck, a conveyor or a concrete pump. The planning of this arrangement must take account of any parking and loading restrictions.

23.4 The size of the site

The way the organisation of the construction work is approached is very different on large sites compared to small ones, even if the quantity of work or size of contract is much the same. A principal object in construction planning is control – control of the work in progress, control of the materials, control of the men working. On a small site, this control can be achieved by carefully siting the foreman's office near to the entrance, within convenient walking distance of the work and concentrating all materials in one storage area, or compound, under the supervision of one store keeper.

Where the site is extensive, this would not prove adequate and may even be counter productive. Several entrances, as may be found on a large site, mean several points to be watched and a gate keeper would need to be appointed. The relieves the foreman of this job and his office can be sited in a position from which access to all the site is more convenient. Placing all the materials at one point may aid security but, on a large site, it means a great deal of travelling to collect the requirements for the work. This can be reduced by arranging several storage areas more conveniently located to the areas of work.

23.5 The size of the contract

Whether the contract is for a single house on a small plot, a whole estate of houses, a multi-storied office block or an industrial development, preplanning of the work and organisation of the construction activities must be carried out. Without it, there would be much abortive work which would have to be taken out to make room for other operations and then re-done. There would also be many delays while waiting for the right materials in the right quantities.

The difference between large and small contract organisation lies in the way in which it is tackled. A small job of familiar content is organised by following an established routine. A regular house builder does not need to start each new commission by carefully analysing the order of working and the organisation to be employed–he will merely follow the pattern successfully used on previous, recent contracts. However, if the project is large, complex and innovative, a very great deal of study of the work must be carried out. The whole operation must be examined by the application of one of the planning techniques, such as the critical path method, and these conclusions transferred to working programmes, bar charts, plant schedules etc., to ensure that labour, materials and plant are brought together at the right time in the right quantities at the right place.

23.6 The type of work

The organisation required for a housing estate comprising a multiplicity of small building projects over a large area is of a very different character to that required for a multi-storied office block. Take the movement of materials, of instance. In an estate, the majority of movements are horizontal from stock pile to work place and, therefore, require haul roads and vehicles or transporters with only a simple means of hoisting up one or two floors at the end of the travel. Where a tall office block is to be built, the main direction of movement is vertically and, therefore, the use of a crane is dictated. A crane is a large, very expensive piece of equipment and consequently the building programme and site organization are arranged to make maximum use of it for the minimum period of time.

The consequences developing from using a crane include its siting where it can pick up and deliver the maximum quantity of goods without further handling: the location of the delivery points, storage points, working areas such as a reinforcement fabrication shop or concrete casting bay within the sweep of the crane, and the location of site facilities such as huts and parking areas, outside the reach of the crane so as to leave the potential storage or working area clear.

Chapter 24

The contractor's legal responsibilities

24.1 Sources of legal liability

A contractor's liabilities at law arise from three sources, criminal law, common law and contractual law. Criminal law is the subject of Parliamentary Acts and to break them is a crime, punishable by a court. Common law covers the acts of individuals and seeks to protect the members of the community against the consequences of those acts. Contractual law is concerned with the relationship between the two people signing a contract, their mutual responsibilities and any consequences arising from their actions.

The contractor's criminal responsibilities are the same as any individual or company. He must abide by the rule of law, pay his taxes and behave in a manner acceptable to our present civilisation.

The contractor's civil liabilities outside those embodied in his building contract are mainly concerned with the law of tort. This can be defined as a civil wrong–i.e. an offence against an individual–as opposed to a criminal wrong or an offence against the state, and its practical consequences are concerned with the adjustment of losses or compensation for damage suffered by the person wronged. There are several forms of tort: negligence is by far the most important, but nuisance, trespass and vicarious liability are also torts.

In both criminal and civil law, the alleged offence must be judged against existing legislation and past cases to establish whether an offence has been committed, but with contract law much of the consequential action and damages arising out of failure to meet contractual conditions is defined in the contract and does not need to be referred to an arbitration or court. Not all claims can be settled by direct reference to the agreed terms of the contract and, in these cases, the dispute is usually referred to an arbitrator.

24.2 Negligence

A firmly established principle in legal matters is the duty of care. This is the principle that anyone, in carrying on their daily life and work, should do so with the exercise of such care for the safety and welfare of others that could be expected to be employed by a 'reasonable' person. This is taken to be someone who weighs up the circumstances, considers the characteristics of the people endangered, takes greatest care when there is greatest danger and never loses his temper.

This duty of care can arise directly or through certain legislation which creates and defines a duty of care for specific people. An example of the direct duty was when a building control officer (and through him the local authority) was held to owe a duty of care when inspecting foundations to ensure that they complied with the Building Regulations. At the same time, the builder owed a duty to make certain that his work met the standards laid down. In both cases, the duty was owed to the ultimate purchaser of the house whose property suffered damage as a result of the negligence of both these people and who was able, successfully, to claim compensation.

There are two Acts of Parliament in particular which create and define duties of care related to a builder's activities: the Occupiers Liability Act 1957 and the Defective Premises Act 1972.

When a contractor takes possession of a site for the purpose of carrying out his contract, he becomes the legal occupier of the property in exactly the same way as someone renting a house and, as such, carries responsibilities under the Occupiers Liability Act. These responsibilities can be summarised as owing any visitor, lawfully on the premises, a duty to take reasonable care so that the visitor will be reasonably safe in using the premises for the purpose for which he is permitted or invited to be there.

This duty only extends to someone lawfully on the premises by licence or invitation, i.e. someone who is allowed to enter the site or asked to call. It does not impose any responsibilities towards trespassers–although a special case would probably be made if a child trespasser was injured due to the contractor's negligence–but this cannot be turned the other way round, permitting the builder to leave parts of his site in a deliberately dangerous condition to deter or trap trespassers.

To be absolutely clear of any claims in this respect, the contractor should:

1. display suitable notices such as: Trespassers will be prosecuted; Warning–Keep out, dangerous; Warning–Dangerous structure;
2. erect and maintain secure fences to deter possible trespassers;
3. eliminate dangerous situations or make them safe: immobilise plant and vehicles, fence or cover excavations, remove or plank up ladders, do not leave dangerous structures unsupported (this applies particularly to demolition work) etc.;
4. actively discourage children from coming on to the site. Building sites are a great attraction to children and the contractor would be held liable for any accidents even if they are trespassing. The younger the child, the greater the liability.

When the lawful visitor is a workman employed to carry out a particular task, he can be expected to guard against the special dangers of his trade. Thus, if a workman is asphyxiated by fumes, of which he was warned but against which he took no protection, then the contractor would not be held to be liable. Nonetheless, the contractor, as an employer, owes a duty of care to his employees to see that the site and equipment on it are reasonably safe to use for the purpose intended.

The contractor's and his employee's duties in respect of safety on site are set out in the various Construction Regulations and in the Health and Safety At Work etc. Act 1974.

The duty of care also extends to the general public and adjoining owners in situations where the building operations are adjacent to a public highway or space and, in particular, where demolition works are involved. Scaffolding alongside or even over the footpath must be specially constructed to ensure nothing falls on to the heads of passers-by. Fans should be erected (see Fig. 24.1) or the scaffold completely enclosed by sheeting to prevent anything falling off the edge of the working platforms, and the platforms should be double-planked with a layer of polythene sheet between them to prevent dust and grit falling through. Where the scaffolding is over or obstructs a footpath, it should be painted white and fitted with red warning lights. All tubes, tube ends and any other potentially dangerous points should be well padded. In deciding what should be done to make such obstructions safe, the contractor should bear in mind the hazards which they present to blind people and the temptations offered to children.

Where demolition or construction work is to be carried on close to the boundary, any adjoining owners have the right to expect the contractor to exercise reasonable care to ensure that their property is not damaged in consequence. To this end it may be necessary for the contractor to provide temporary support, in the form of shoring to the adjoining property, and temporary weather protection with

256

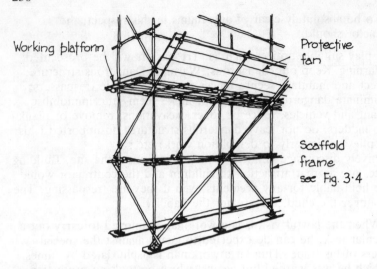

Working platform

Protective fan

Scaffold frame see Fig. 3·4

Fig. 24.1 Protective scaffolding fan

polythene or tarpaulin sheets until the new work is completed. If there is a possibility of damage occuring, it is usual and prudent to carry out a survey of the condition of the adjoining buildings, in the company of the owner or his representative, to record any faults existing before the work starts and, if permitted, to secure a glass tell-tale across any cracks to indicate whether any further movement occurs during the building work. A tell-tale is a narrow strip of glass placed across the crack and securely fixed to the wall each side. Any movement occurring is clearly indicated by the glass breaking.

24.3 Nuisance

An ever-present hazard in the carrying out of building work is the creation of a nuisance. For this to be actionable by the person or persons suffering damage, the nuisance must be repeated and must lead to loss. A man with a pneumatic drill working late at night on one isolated occasion, disturbing the sleep of adjoining residents, is not likely to lead to a claim being upheld in court, but if the offence was repeated regularly and the health of the neighbours began to suffer, then the court might well take a more serious view.

There are two types of nuisance, public nuisance and private nuisance. Public nuisance is an offence against the state and is a punishable crime; typical examples are obstructing public highways by digging holes, depositing materials or leaving a skip in the road without making proper arrangements, or obtaining an appropriate licence from the local authority, emitting an excessive amount of smoke, dust or fumes, creating an excessive amount of vibration and noise.

Under the Control of Pollution Act 1974, local authorities were given the powers to control noise on construction sites. They can do so by serving a notice restricting the use of specified plant, limiting the hours of work and controlling the upper level of noise. Also, under this Act, the contractor may, and if there is a risk of nuisance should, obtain the prior consent of the authority to his proposed methods of working. Other relevant Acts in relation to public nuisance are: the Clean Air Acts 1959 and 1968; the Highways Acts 1959, 1971 and 1972; and the Public Health Acts 1936 and 1961, much of which has been transferred to the Health and Safety at Work etc. Act 1974.

Private nuisance can be defined as the unlawful interference with the use or enjoyment of another person's land. This applies to all who have possession of land and, therefore, it is possible for an adjoining owner to commit a nuisance to a contractor by interfering with or obstructing his work, but it is more usual for any actions to arise from the contractor's activities causing a nuisance to local people. To minimise this risk, the contractor should take steps to limit the emission of smoke, dust or fumes, and the creation of vibration and noise and to avoid the obstruction of entrances by careless parking, flooding or digging of trenches. Where the last of these must be carried out, temporary arrangements must be made to allow adjoining owners to enter and leave their premises, usually by placing a heavy steel plate across the trench.

As well as interference with the use of the land, nuisance can arise as a result of interference with certain rights over land. Such rights are established when an adjoining owner gains a benefit from the area in which the contractor is working. The more common are right of support, right of light, right of air, water rights and rights of way. The right of support can be violated by digging away the subsoil supporting the foundations of the adjoining property (see Fig. 24.2) thereby causing subsidence. Right of light and air, and rights of way would be infringed by the building design, which must take these into account.

24.4 Trespass

Trespass by members of the public is dealt with in section 24.2, but the contractor may also be held to have committed one of the forms of trespass, unless his site staff are aware of the law. There are three forms of trespass:

- trespass to persons

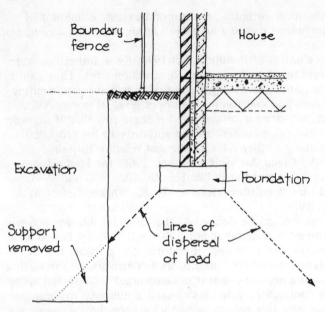

Fig. 24.2 Loss of right of support

- trespass to goods
- trespass to land

The last is the one most commonly understood but either of the other two may be committed during building work.

Trespass to persons arises in cases of assault, battery or false imprisonment. Assault is the attempt or threat to apply force to another person unlawfully. This could arise on a site if the foreman has to deal with a particularly persistent intruder and, for example, tells him to 'get off this site or I'll bury you in the concrete'. Such a threatening statement is assault. Battery is the action of applying force–pushing, hitting or throwing something at another person–even though no physical injury or damage results. For example, if the threat expressed above had no effect and the foreman took hold of the intruder's collar and propelled him towards the wet concrete to scare him off–this would be battery. If, subsequently, the scare worked and the intruder ran off but, to reinforce his message, the foreman threw a brick after him, this would also constitute battery.

Alternatively, if the intruder did not respond to the initial threat and the foreman said 'Right–stay there until I fetch the police' and, to ensure that the intruder did not move, stationed two labourers armed with pick-axe handles as guards, this would be false imprisonment, unless the intruder had been caught 'red-handed' stealing or there was evidence of forceful trespass.

In all cases, no matter what the provocation, it is prudent to deal with difficult situations with patience and tact, and without recourse to violence.

Trespass to goods arises where there has been a forceful interference with the possessions of other people. This is more likely to arise in contracts where alterations to existing premises is involved rather than new building works and, to guard against any claims arising in this respect, the contractor's staff should be warned to leave alone any of the building owner's property that does not concern them, no matter how innocent the inquisitiveness.

While both the foregoing arise through the unlawful action of the contractor's employees, trespass to land will usually arise by the execution of the work. This trespass may take place on the surface of the land, under the surface or in the air space above the surface and may constitute entry on to, remaining upon, or throwing or placing objects upon or in the land.

The most commonly recognised form of trespass, walking across the land of someone else, must be avoided by the workmen engaged on the building unless a prior agreement has been reached. This can easily arise where the new building is on or very near to the boundary and the workman steps over the fence to gain access to his work.

There are a number of other trespasses to land which can occur when the building is on the boundary. Take strip foundations, for example. The wall must be built along the centre line of a strip foundation but, if the face of the wall is the boundary, the projecting part of the foundation is over the boundary and is trespassing on to the adjoining land, Similarly, if the wall is finished with projecting eaves or coping, this also would be a trespass; likewise any other projections (see Fig. 24.3). Furthermore if, to build the wall, normal scaffolding is erected, this must be on the land of the adjoining owner and would constitute a trespass unless it was with his agreement. Nor is it necessary for the scaffold to stand on the surface of the land to cause an offence. A scaffold frame cantilevered out from the building at an upper level and overhanging the boundary still constitutes trespass. This trespass into the air space above the land also occurs when a tower crane is carelessly sited or used so that its rotating jib passes over another person's air space.

The permanent offence caused by the incorrect design of foundation eaves and other projections must be avoided at the design stage, but the aggravation caused by thoughtless trespass during the construction period can easily be avoided by an advance meeting with the adjoining owner to seek to obtain his agreement to the proposed method of working.

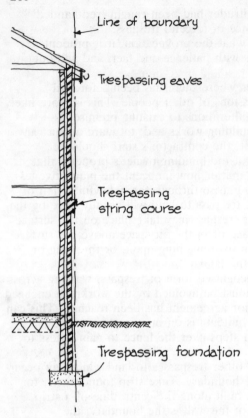

Line of boundary

Trespassing eaves

Trespassing
string course

Trespassing foundation

Fig. 24.3 Trespass

24.5 Vicarious liability

Vicarious liability is, briefly, a liability for the torts of others. The building owner may be liable for the torts of the contractor, and the contractor for those of his sub-contractors, and in all cases the employer may be liable for the torts of his employees.

There are many exceptions to this rule; for instance, where the contractor is considered to be independent rather than a handyman. A handyman is told what is to be done and, in general, how he is to do it, whereas an independent contractor is merely told what work is required and assumes the responsibility for deciding how it should be done and for carrying it out in a diligent and workmanlike manner. The same rule extends to the responsibilities between a contractor and an independent sub-contractor.

Even though the employer takes on an independent contractor and the contractor employs an independent sub-contractor, each may be liable for negligence in selecting a contractor or sub-contractor possessing sufficient competence to carry out the work or for failing to give sufficient directions on the execution of the work.

When the law deals with vicarious liability between employer and employee, it uses the words 'master' and 'servant' and endows them with a rather wider meaning than 'employer' and 'employee' to cover anybody hired to carry out work other than as an independent contractor. In this area of legislation, the master is vicariously liable for the authorised and unauthorised torts of his servant provided that the latter were committed within the course of his employment.

Thus the master is liable if his servant carries out the work he is employed to do in a negligent manner or even in a way he was expressly forbidden to do, but not if he exceeds the scope of his employment. This was shown in the case of Conway v. Wimpey in 1951 when there was an accident involving a vehicle carrying an employee of another firm on the site. The driver was held to be negligent, but since he was employed to carry fellow-employees only, his act was held to be one for which he was not employed and his employer was not liable.

Whether or not the master is liable, the servant is generally liable for his own torts and may be sued separately or jointly with the master. Bearing in mind that damages may only be recovered once and that in a joint action one defendant may have more money than the other, it is important that the plaintiff selects those to be sued who are likely to be able to satisfy a judgement.

24.6 Contractual responsibilities

As already mentioned, the formal building contract sets out, among other matters, the responsibilities the employer, i.e. the person having the work done, and the contractor bear to each other and the remedies immediately available should either fail to fulfil their obligations.

With regard to the contractor's responsibility towards the general public, the contract makes the contractor liable for any third-party claims in respect of death or personal injury, or damage to property arising out of the carrying out of the work. It further makes it obligatory for the contractor to take out adequate insurances to guarantee payment of his liabilities, thus ensuring that the employer is fully indemnified against any claims which otherwise may be passed on.

The contractual responsibilities commence with the date for possession of the site stated in the contract–not necessarily the date when the contractor starts work on the site–and ends on the date shown on the Certificate of Practical Completion which is issued by the architect when the work is complete for all practical purposes.

Chapter 25

Materials, delivery and storage

25.1 The need for control

There is a history – almost a tradition – of wastage of building materials. It probably originated when buildings were constructed from materials, either new or second-hand, which were convenient to the site and any wastage was easily made up by helping oneself to another log in the forest or a few more stones from the local ruined castle.

Today, very large quantities of materials, originally intended for a particular building, never find their way into it. The deficiency is largely accounted for by wastage, through a variety of causes, and the rest disappears, either to be used for other purposes on-site or completely off the site altogether, stolen by amateur or professional thieves.

The Building Research Establishment carried out a survey of materials wastage and reported in 1976 in Current Paper CP/44/74, *Wastage of Materials on building sites*, that between 10 and 20 per cent of all materials delivered to site ended up as waste or was illegally removed during the contract. The enormity of the problem is only appreciated if these percentages are translated into figures. Taking the average of 15 per cent loss and applying it to the construction statistics published for 1983, one finds, for instance, that something over 500 million bricks disappeared! – this would be enough to build the outer skin of the external cavity walls of some 50 000 houses! In the same area of craft, 15 per cent loss of lightweight concrete blocks amounts to nearly 4 million square metres or enough for the inner skin and internal partitions in all 50 000 houses! Looking at the financial loss, a deficit of 15 per cent of timber used in 1983 amounts to just over £10 m. and for plywood just over £2.5 m.

It is clear that there is room for a considerable improvement in the way we use building materials but, unfortunately, the situation can arise where it costs more to prevent the losses than the value of the material lost. In that case, many builders simply order enough to allow for the loss to occur without causing inconvenience on-site due to a shortage. This is an easy answer to the problem but one which, inevitably, puts up the cost of building and one which merits close examination at regular intervals.

25.2　Materials, ordering and delivery

The materials for a building contract are not all delivered at the beginning of the construction period. This would place an unnecessary strain on the manufacturer's production resources, on the contractor's financial resources, and on the in-situ storage facilities. It would also increase the risk of loss through damage and vandalism. The usual arrangement is to phase deliveries over the length of the construction period so that the requisite material arrives just before it is needed.

Ordering of the materials should all be done early in the building programme or may even have been placed on a provisional basis, by the architect, before the contract is let. When the bulk order is placed, it would be accompanied by a call-off schedule setting out when each delivery is to be made and how much is required in each load. Take, for example, a small estate of a dozen houses programmed to be completed in a year. Bricks would be needed for the foundation walls – say 4000 per house – for the external skin of the cavity walls – 10 000 per house – and for external works (garages and walls) – say 2000 per house: giving a total of 192 000 bricks. These would be ordered – some for foundation work and some for facing work – as soon as the builder received confirmation of his contract. But in the first month he would only need the bricks for the foundations to, say, three houses, i.e. 12 000, then in the second month he might well bring in more bricklayers so that he can continue with the first three superstructures while commencing on three more foundations which would need another 12 000 foundation bricks plus 30 000 facing bricks.

The pattern could be repeated in the third and fourth month but with the addition of the external works calling for 6000 bricks each month, so that the total delivery would be 48 000 bricks. By the end of the fourth month, the foundations would be finished and the order would reduce by 12 000 in month five. In month six the superstructures would have been finished so all that would be needed would be 6000 bricks for the three remaining garages and boundary walls. After the six months, no more bricks would be needed.

Three main reasons why an architect may pre-order or reserve materials in advance of the contract being signed are, firstly, that the delivery period is so long that it would not fit into a construction programme starting with the signing of the contract–this is particularly critical with materials required early in the job; secondly, that part of the work was designed by a specialist, e.g. the steel frame, and he demanded a letter of intent to order before he would agree to carry out the detailed design work; thirdly, that the work is to be carried out by one of the statutory undertakers such as the Water Board laying a water main and they need advance warning to fit it into their programme.

Having ordered the materials, preparations must be made on-site to receive them. The nature of these preparations depends on the value of the goods, its sensitivity to weather, the amount and the method of despatch.

Various storage arrangements for materials and goods of differing value, sensitivity and bulk are dealt with in the next section. Methods of despatch vary widely and have developed new features in line with more sophisticated handling equipment. Broadly, there are five forms in which materials or goods will arrive, bulk, bagged, palleted, packaged or loose.

Materials delivered in bulk are mainly topsoil, hardcore, sand, aggregate, cement and ready-mixed concrete. The first four are off-loaded directly from the lorry into a storage area or container by tipping. It must be possible, therefore, for the lorry to drive as far as the aggregate store in a fully-laden condition. Cement in bulk is delivered in a sealed container lorry and blown into a specially designed cement silo for use in a centralised concrete batching plant. Again, the delivery vehicle must be able to reach the silo. Ready-mixed concrete can be deposited in an arc within 3 to 4 m of the discharge end of the mixing drum on the lorry by means of pivoted chutes. If this is insufficient to reach the point where the concrete is required, it is either discharged in small batches into dumper trucks which transport it, or into a conveyor or concrete placer which carries or pumps it to the desired destination.

To ease handling, bulk materials such as cement are placed in bags sufficiently small to be manhandled. Being small, the speedy unloading of a full lorry-load generally requires the employment of several men.

The handling of bagged materials and some materials like blocks, which are also delivered loose, is greatly simplified by placing them on a pallet. This is a wooden frame, rather like a shallow box without ends, into which the forks of a fork-lift truck can be inserted. Needless to say that, to take advantage of palletised loads, there must be a fork-lift truck available on the site.

An alternative to palleting loads is to package them. In the past this has always been the system for small components such as ironmongery but now concrete blocks and bricks are delivered either banded together or enclosed in a shrink-wrapped clear plastic package; roof trusses and drain pipes are banded; many components such as kitchen units and sanitaryware are in a ready-to-assemble form and boxed etc.

Delivery of materials in a loose form is less common now than it used to be. Roof tiles are still delivered loose by some manufacturers but the delivery lorry has a specially designed on-board hydraulic grab which can pick up ten or twelve tiles at a time and stack them neatly beside the lorry. The only other main class of components delivered loose is specially made elements, particularly joinery items, which tend to be unsuitable for palleting or packaging.

25.3 Materials and component storage areas

On all sites, careful consideration must be given to where materials and components will be stored. Where space is restricted, the location of the storage area is probably pre-determined, but where there is plenty of room, control must be exercised to prevent stored materials being unduly spread out thereby making the handling less efficient and increasing the risk of damage or disappearance. For example, take a banded package of bricks permitted to be off-loaded on a convenient but uneven area of land just beside the temporary haul road instead of in a properly levelled and prepared brick storage area. The package would be safe, standing at a slope, until the bands are cut. Once this is done, to obtain the bricks, there is a good chance that some would fall off the stack on to the road. The next time a lorry passes, these bricks are then crushed down into the road – and a few more join the 500 m. annual loss.

One decision to be made by the contractor is whether to have one large materials compound or several small storage areas. The single compound is easier to control, more secure and generally is chosen when the contract involves the use of a crane. On some contracts, such as a housing estate, a single compound could result in an excessive amount of on-site transport simply carrying the materials from where they are stored to where they are needed. In this case, several storage areas would be planned to follow the order of working round the site: as each phase is completed, the earlier storage areas are discontinued.

In some cases it may be planned to have a number of general storage compounds around the site and a single static area for, say, on-site concrete mixing, since a concrete-mixing plant and its attendant cement silo and aggregate bins is quite an expensive installation to move, or an on-site joinery or reinforcement fabrication shop which would also be uneconomic to relocate.

The final decisions will rest on considerations of point of use, ease of access and off-loading, rate and sequence of working, type and quantity of materials. A primary consideration is to place the delivered material as close to its point of use as possible and, if that is at a height, the store must be convenient to a lifting appliance. Most suppliers make it a condition of sale that their vehicles will be provided with a safe and easy means of access, which means a well-consolidated road, either the base course of a permanent road or a temporary road of sleepers. Nor should the delivery vehicle have to leave the road to unload; therefore, the store must be adjacent to the road.

The rate and sequence of working may, as already discussed, lead to several, successive, storage areas. It can also mean that one area may be used for a succession of materials, i.e. the brick storage area could be used for roof tiles once the walls are up, then for drain pipes when the tiles have been used and, finally, for fence posts and pavings after the drains have been laid. The type and quantity of materials dictates the method of storage, which is dealt with in the next section, and will also determine the form and location of the storage area. For instance, aggregates would be stored in the open near the concrete mixer; plumbing fittings are often stored in a room in the partially completed building.

25.4 Methods of storage

All building materials require particular provisions for their efficient storage to ensure that they are maintained in a satisfactory condition until required and are not subject to loss through damage or pilfering. Following is a list of some of the more common groups of materials which indicates their special requirements for storage.

Hardcore: this is delivered in bulk and is tipped in mounds as near to the point of use as possible but not directly into the area to be filled.

Aggregates: all aggregates must be kept separate, in their respective gradings, and tipped on to a prepared base, preferably of concrete laid to falls. Different aggregates can be divided and contained by vertical divisions to form storage bins, thus reducing the area required.

No aggregate should be tipped directly on to the surface of the soil as this leads to possible contamination of the material by soil water chemicals and wastage of the lowest layer of the heap.

Cement: In bulk form, cement is stored in a specially designed silo. In bags it must always be stored on a dry platform clear of the ground, protected from the weather (preferably in a shed), no more than 2 m high and so arranged that the bags can be used in the order of delivery, i.e. the first delivery used first (there is a tendency for subsequent deliveries to be placed on top of earlier bags, thereby delaying their use until eventually the cement is no longer usable).

Bricks and blocks: Banded or packaged loads should be neatly stacked not more than three packs high on a prepared level base where there is no risk of mud splashing up the stack.

If delivered loose, they must be carefully stacked with sloping ends to the stack so that there is no danger of falling off (see Fig. 25.1).

Fig. 25.1 Stacking bricks on site

Generally, these are satisfactory stored in the open but the stack should be covered if threatened with excessive rain, snow or frost.

Roof tiles: Stored in rows on end on a clean flat surface. The rows are retained by a few tiles stacked flat at the ends.

The storage area should avoid situations where the tiles can be covered with mud or concrete splashings.

Carcassing timber:	This is the large section sawn timber used for floor joists and structural frames and should be stacked level on an even platform well clear of the ground in layers separated by spacers to ensure a flow of air round the timber. The stack should be covered against rain but the ends left open for ventilation. Timbers should be stacked according to type and cross-sectional size.
Roof trusses:	Trussed rafters arrive in banded lots and should be stacked either horizontally with spacers between each pack or vertically on a rack (see Fig. 25.2). In both cases, a clean firm platform is needed and the stack should be protected from the weather.
Joinery items:	The timber of frames, doors, windows, staircases etc. is usually drier than that used for carcassing and, therefore, must be stored under cover, preferably in a shed but allowing good ventilation. The items should be stored flat and adequately supported to avoid straining any joints. Doorsets (door frame and door complete) should be stored 'pull' side up so that the door rests on its stops.
Boards:	Incorrectly stored, plywood, chipboard, blockboard and hardboard can acquire a permanent bow which makes it difficult or impossible to achieve a satisfactory job. They should never be stored on edge, but always flat with adequate support and spacers at intervals. The stack must be well protected from the weather, preferably in a shed, and good ventilation provided.
Roofing felt:	Rolls of roofing felt should be stored on end, as indicated on the roll, on a clean, dry level surface. They need to be shielded from the effects of hot sunshine.
Plaster	Plaster is always in bags and should be stored in the same way as cement in bags. As the building should be weathertight before plastering commences, the bags of plaster are often stored in one of the rooms.
Plasterboard:	Plasterboard is easily damaged and must be stacked horizontally on a level dry fully-supporting platform not more than 900 mm high in a shed or completely sheeted over if

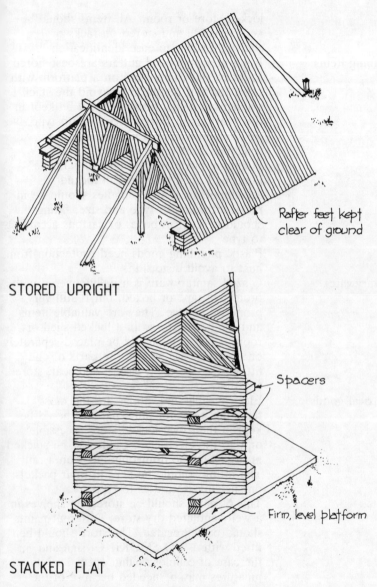

STORED UPRIGHT

Rafter feet kept
clear of ground

Spacers

Firm, level platform

STACKED FLAT

Fig. 25.2 Storing trussed rafters

	outside. If the shed has a concrete floor, the boards should be laid on a polythene sheet to keep them dry.
Ironmongery:	All ironmongery items need careful control and should be kept secure, preferably in a

	locked store or room. All items should be kept separate and stored on racks or shelving in bins to ease identification.
Plumbing items:	Rainwater pipes and gutters are best stored in racks, but can be laid on a platform with some form of retention to avoid the stack spilling off. Small fittings should be kept in their cartons or boxed according to type. No weather protection is needed except shielding from very hot sun.
	Plumbing pipes and fittings are valuable items and must be stored in locked accommodation. Pipe bundles should be in racks according to type and size, fittings should be kept in boxes or cartons according to type.
	Plastic plumbing goods need protection from heat to avoid distortion.
Sanitaryware:	Glazed sanitaryware is usually covered with protective tape or boxed, but is still very prone to damage. They are valuable items and should be stored in a locked shed or room. Each item should be placed separately on an even base and all the parts of one fitting, i.e. WC pan, cistern and seat, stored together.
Electrical goods:	Electric cable and fittings are the most tempting items for pilfering and should be kept in their own locked store accessible only to the electrician. Cable can be stacked in drums or the drums mounted on a rail, The fittings should be kept in their packets, on shelving, sorted as to type.
Paints:	Tins of paint should be stored on shelves in a well-ventilated dry store. Different paints should be segregated. The store should be fitted with a 'NO SMOKING' sign and, in the case of cellulose paint, special safety measures may be needed because of the serious fire risk.

25.5 Reduction of waste

There are a number of causes of waste, not necessarily confined to malpractice on site. It can start on the drawing board by the designer creating a building which generates waste by failing to suit the standard sizes of the constituent materials. It is impossible to design every part of a building to fit every type of material. To achieve this the designer would, for example, need to select room widths which fitted multiples of partition blocks, multiples of plasterboard ceiling and possibly wall lining, stock lengths of skirting board (after deducting twice the plaster thickness) and even multiples of wallpaper widths (also allowing for plaster on the returns) or stock carpet sizes, to fit between the skirtings, and this is ignoring the size the room needs to be for its intended function. Nonetheless, if a brick building is planned, then the dimensions should be multiples of the brick module and floor-to-ceiling heights can be selected to suit partitioning units etc.

A second cause of waste is in transport: incorrect loading or inadequate packaging can lead to goods being damaged when they reach the site.

On-site wastage occurs through misinterpretation of the drawings and deliberate over-estimation of quantities required to avoid running short, mis-use, faulty workmanship, careless handling and storage, deliberate or mischievious vandalism and pilfering.

Calculating the precise quantity of materials needed is a difficult, often tedious, task which, if incorrectly done, can lead to delay if insufficient materials are ordered and an unacceptable job if the make-up quantities do not match the first load precisely. Furthermore, the builder is often pushed into over-ordering because what he needs is not an economic quantity. A nearly full lorry-load of bricks can cost more than a full one because of the extra handling involved in unloading a part load at the builders merchant's yard between the brick works and the site.

Mis-use of materials can be extensive on a badly supervised site: bricks used for all number of purposes other than building walls, tile battens used as pegs, temporary supports etc. Faulty workmanship giving rise to work having to be repeated is also a consequence of lack of proper supervision. Careless handling and storage is probably the most fruitful source of waste and one very easily avoided by careful planning and proper training. Deliberate vandalism (adults rampaging round the site purposefully destroying the work) and mischievous vandalism (children playing with the sand and mixing stones or other materials with it) should be controlled by security measures. Pilfering, if professionally carried out, is difficult to counter–the stories of thieves stealing large excavators or arriving on-site suitably dressed and being helped to steal large quantities of materials are almost unbelievable. The amateur thief (sometimes one of the builder's own employees) who only wants a few bricks to build a base for his greenhouse is easier to catch and represents a steady siphoning-off of the builder's stock of materials unless checked.

25.6 Site security

At one extreme, site security can mean high fences, barbed wire, patrolmen, guard dogs, floodlighting and automatic burglar alarms. At the other end of the scale, it is a locked cupboard in the site office for the really valuable or dangerous items. The optimum level to be found between these two depends on many factors concerned with the contract, its size and value, the type of work and the location of the site.

There are a variety of security measures which can be employed either singly or in combination; the following is a review of the most commonly employed:

Fencing:
: This is extensively used and can provide a sufficient deterrent for a lot of intruders, particularly if it is a sheet fence of boards fixed to posts so as to be virtually unclimbable. The height of the fence should be 2 m and any gates must be of equal construction. It is prudent to cut observation panels at intervals in the fence and fill them with grilles, otherwise the solidity of the fence assists a thief, who has managed to scale it, to go about his illegal business unobserved.

Compounds:
: This term describes an area set aside on a site and securely fenced. A variety of materials would be stored within the compound either in the open or within sheds and the whole area is the responsibility of a storeman.

Store sheds:
: Any sheds intended for the storage of materials unsuitable for keeping outside should be lockable, even if they are within a compound. They should be located in groups where surveillance is easy, not in an isolated corner where it is difficult to see a would-be pilferer.

Security patrols:
: Many professional security companies will carry out after-hours unscheduled checks on building sites and the knowledge that such checks are being made–as evidenced by a notice to that effect–will deter many vandals and thieves.

Guard dogs:
: The law now forbids guard dogs to be allowed to run free in a compound and requires that they must be in the constant

charge of a handler. For this reason they are seldom used today.

Night watchmen:
Unfortunately, this is a job frequently given to aged and relatively infirm men, totally unsuited to the job. A night watchman, to be effective, must be young, active, able to make his presence known and felt at all times and fit enough to react with speed to an irregular occurrence. This costs a lot of money and usually the periodic visit by a security company is cheaper.

Floodlighting:
This is a really effective means of combating intruders. It is easy to install and relatively cheap to maintain. To be a complete deterrent, the floodlit area must be constantly and clearly visible from the public highway. The system has the added advantage that it can also be used to provide working illumination during the dark days of winter.

Chapter 26

Site accommodation

26.1 Legislation

The minimum standards of on-site accommodation for the health
and welfare of construction workers was governed by a multiplicity
of provisions before 1974. Most of the legislation was embodied in
the Construction Regulations, which were derived from the Factories
Act 1961, and which were enforced by factory inspectors appointed
under the Act.

In 1974 the Health and Safety at Work etc. Act was passed.
This imposed some important new powers of control and also
incorporated the whole of the Factories Act except Section 135, and
with it the Construction Regulations. The task of enforcing the law
was given to a new body, the Health and Safety Executive operating
under the direction of a Health and Safety Commission, and the
work of the factory inspectors is now transferred to the new health
and safety inspectors with considerably wider powers than their
predecessors.

The intention of the Health and Safety at Work etc. Act is to
repeal or modify the existing statutory provisions which, at the
moment, have been taken over by the Act. (There are many other
parts of or complete Acts besides the Factories Act which are
affected in this way.) At the present time the Construction
Regulations have not been changed, although they will eventually be
replaced, but must now be read as part of the Health and Safety at
Work etc. Act and in the light of the other provisions in that Act.

One of the effects of the Health and Safety at Work etc. Act was to place a duty on both employers and, notably, employees to ensure that all provisions necessary for the exercise of healthy and safe working conditions are adequately available, fully maintained and properly used. This principle is of great significance in ensuring safety on-site but it still applies to the accommodation to be provided for the health and welfare of everybody working on site.

Health and safety inspectors now have the powers to enter premises, to make examinations and investigations, take samples, to take possession of articles or substances, to require persons to answer questions and to require the production of documents. Where a contravention of the provisions and regulations has occurred, the inspector may issue an 'Improvement Notice' which requires changes or improvements to be made within a specified period. If the contravention involves a risk of serious personal injury, he may give the contractor a 'Prohibition Notice', which has the effect of stopping work until the contravention has been discontinued.

There are four sets of Construction Regulations, which are entitled:

The Construction (General Provisions) Regulations 1961
The Construction (Lifting Operations) Regulations 1961
The Construction (Working Places) Regulations 1966
The Construction (Health and Welfare) Regulations 1966

It is the last of these which controls the various facilities to be provided on site in the interests of health, first aid, shelter, welfare and sanitary accommodation.

26.2 Accommodation for site staff

In addition to the requirements of the Construction Regulations, certain accommodation must be provided for administrative purposes. The variety and extent of such office accommodation varies according to the size of the contract, the complexity of the work and the administrative organisation.

Apart from very small building operations, lasting a matter of days only, all sites will always have, as a minimum, an office for the resident site manager, usually a foreman. This is required to provide him with somewhere in which to keep all his records, receipts and registers, the drawings, specifications and bills of quantities relating to the work and any other documents essential to his administrative function. It also provides him with a place in which to work when carrying out that function. It should be sited where the occupant can maintain observation of the entrance to the site and the work in progress.

On large contracts, this foreman's hut can develop into a large, sometimes two-storied, prefabricated block of offices accommodating site managers, site surveyors, site engineers, site architects, general foremen, trades foremen, finishing foremen, bonus surveyor, timekeeper, clerk of works, sub-contractor's on-site staff and secretarial support staff.

It is usual for the main contractor to provide the site offices both for his own use and for use by the clerk of works and any site architects or engineers or sub-contractor's staff. The cost of this accommodation is borne by the client and is included as a preliminary item in the builder's tender.

Precisely what type of temporary structure is used and whether it is owned or hired by the contractor is a matter for decision at the pre-tender stage, taking into account the nature and duration of the project and the builder's own financial policy with regard to capital expenditure on such items. The types of temporary accommodation available can be divided into sectional timber or glass-fibre buildings, one-piece cabins or mobile caravans.

Because of their ease of storage and transport, the sectional panel-type buildings are usually favoured by builders who wish to own their site buildings. These can be readily manhandled and carried, in dismantled form, in a normal lorry and, when not in use, individual panels do not occupy a lot of space in the builder's yard.

It is also possible to combine a variety of panels to obtain differing sizes of hut and alternative door and window arrangements according to the demands of the site. Inevitably, the repeated assembling, disassembling and transporting leads to accelerated wear and damage and, therefore, the useful life of these buildings is relatively short.

The sectional buildings tend to be single-storey sheds, frequently with a pitched and felted roof. The one-piece cabins are of a rectangular cell form and can be stacked on top of each other if a multi-storey office structure is required (see Fig. 26.1). They usually require a special transporter because of their size and may need a crane for off-loading or multi-storey stacking, although some manufacturers provide a four-point jacking system, as shown in Fig. 26.1, so that they can be off-loaded by the lorry driver and can stand level on uneven sites. Larger firms of builders may find that their demand for site accommodation is such that there is a financial advantage in owning their own cabins of this sort. They tend to last longer than the sectional buildings and the problem of space taken up by storage is solved because the demand is such that they are taken directly from one site to another. For other contractors, with less of a continuous demand for this type of office, hiring them from specialised companies for specific contracts is an economic alternative.

The mobile caravan, similar to a normal holiday touring caravan

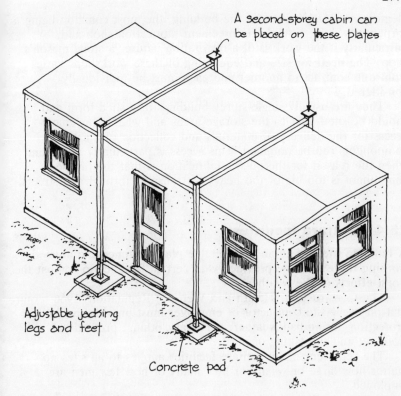

A second-storey cabin can be placed on these plates

Adjustable jacking legs and feet

Concrete pad

Fig. 26.1   One-piece cabin

but fitted up with office facilities, is very useful for builders specialising in smaller contracts where the foreman's office is frequently moved to each new job; or for contracts, such as housing estates, where the site accommodation may be moved several times as each phase of the project is completed.

As well as providing the office building, it is also necessary to fit it out with a desk, or desks, with lockable drawers, tables for laying out drawings, drawing boards, chairs, stools, lockable cupboards, plan storage arrangements – either a plan chest or suspension plan files – book shelves, a telephone and heating facilities. Periodic office-cleaning arrangements should be made and the tender must include an allowance for clearing the whole lot away when the job is finished.

26.3 Workshops

On many sites, workshops are provided for on-site fabrication of elements or components for the building, the most common being a carpenter's shop, a steel reinforcement fabrication shop and, particularly if the work is of a restoration nature, a stone mason's shop. The need for size and equipping of these workshops varies from one contract to another and each must be individually considered.

They are usually single-storey buildings of a shed form and should be sited near to the storage areas and where convenient access for the delivery of materials and collection of finished components can be obtained. This access is particularly important when the reason for the on-site fabrication is that the finished component is too big or too heavy to transport through the streets.

26.4 Welfare facilities

A building site can be an inhospitable place at times and anybody working there must be provided with certain facilities to combat the worst effects of the weather.

The Construction (Health and Welfare) Regulations 1966 require that on every site contractor or employer must provide shelter for: protection during bad weather; personal clothing; protective clothing; and the taking of meals.

The requirement for welfare facilities applies to all sites no matter how large or small nor how many or how few men are employed.

On most building sites it is normal for more than one employer to be engaged on the work: there is usually a main contractor and a number of sub-contractors. It is not necessary for each to provide his own accommodation as this clearly would be uneconomic but, if accommodation is shared, the firm providing the facilities (usually the main contractor) must allow enough room for the total of its own employees plus those of the firms sharing the accommodation. Whoever provides the shelters, mess huts etc. must also keep a register showing what facilities it is proposed to share and the names of the firms sharing them.

Where more than five people will be employed on the site, the shelters for both the employees and their clothing must be warmed and there must be arrangements made for drying clothing. If fewer than five are employed, shelters must have such arrangements for warming and drying as are practicable.

On sites where the work force exceeds ten, the mess room must be provided with the means of heating food, but some means of boiling water must be provided on all sites, no matter what number are employed. Every site must also be provided with drinking water at convenient points and these must be clearly labelled 'DRINKING WATER'.

The accommodation provided is usually of a type similar to that selected for the site offices, described in section 26.2, and the size needed can be found by allowing 2.0 to 2.5 m² per person for the mess room and 0.6 m² per person for a drying room in addition. Tables, chairs and cupboards for utensils should be provided in adequate numbers for the people employed, the interior must be kept clean and in good repair, and it must not be used for storing materials. This site accommodation should be positioned well clear of the building operations and the storage, delivery and workshop areas. Wherever the temporary buildings are placed, they must be reached by proper paths which are maintained in good order.

The facility for heating food in the mess room is not required if there are canteen facilities on the site. Such canteen provisions are subject to the Food Hygiene (General) Regulations 1970, the basic purpose of which is to prevent illness developing from contaminated food. The responsibility for enforcing these regulations lies with the local environmental health officer. The requirements specify that:

Articles or equipment with which food is likely to come into contact must be non-absorbent and completely cleansable, i.e. smooth metal, laminated plastic or plastic-coated board work-surfaces and wall linings, and they must be kept in good condition.
People handling food must be clean, must not spit or smoke, must cover cuts with a waterproof dressing and must wear clean over-clothing and neck and head covering.
An adequate supply of hand-washing facilities must be provided, separate from any vegetable- or dish-washing sinks.
The room must not communicate directly with sanitary accommodation. The room must be adequately lit and ventilated.
The walls, floor, ceiling and woodwork must be kept clean and in good repair.
No waste material must be allowed to accumulate.

26.5 Washing and sanitary facilities

Every site where anybody is employed for more than four hours must be provided with washing facilities of an adequate and suitable nature. On sites where more than twenty men are employed or where the work will last for more than six weeks, these washing facilities must include: adequate troughs, basins or buckets with smooth impervious internal surfaces; soap and towels or hand-driers; a sufficient supply of hot and cold or warm water.

On larger sites where more than 100 men are working and the work will last for at least a year, the Regulations specify a minimum of four washbasins plus one extra for every additional thirty-five men, i.e. 100 to 135 men – 5 basins, 136 to 170 – 6 basins and so on.

Where any men handle poisonous substances, i.e. lead and certain chemicals now used, the washing facilities must be increased to one basin for every five men and nailbrushes must be provided in addition to the soap and towels or hand-driers and hot and cold or warm water.

The washing facilities provided must be near any mess rooms and must be kept clean.

Sanitary conveniences, i.e. water closets or chemical closets, must be provided at the rate of one for every twenty-five persons employed on the site with separate facilities for men and women. Where adequate urinal fittings are installed, the rate of provision of sanitary conveniences over four, i.e. more than 100 employed on the site, can be reduced to one for every thirty-five persons. Every sanitary convenience, not a urinal, must be under cover, screened to ensure privacy, provided with a proper door and fastening, ventilated and properly lit. Urinals are merely required to be placed or screened so as not to be visible from anywhere on or off the site. All facilities must be conveniently accessible at all times but must not open directly off a mess room, work room or canteen and must be kept clean.

26.6 First-aid provisions

Formerly, the requirements to provide first-aid facilities were to be found in the Factories Act of 1961, the Offices, Shops and Railway Premises Act of 1963 and the Construction (Health and Welfare) Regulations 1966. These were repeated or revoked when the Health and Safety (First Aid) Regulations 1981 came into effect in July 1982, accompanied by an Approved Code of Practice and Guidance Notes.

The regulations place a general duty on employers to see that adequate first-aid provisions are available for anybody who may be injured or become ill at work; the Code of Practice defines the provision to be made; and the Guidance Notes advise on equipment, facilities and training.

In Regulation 3, employers are required to ensure that there is a sufficient number of suitably qualified persons who can render first aid. For most building sites, this means one trained first aider normally present when the work force numbers between 50 and 150, plus additional first aiders for every 150 additional workers. As for welfare facilities, the first-aid provision may be shared between employers on the same site and similarly registered. Where there are fewer than fifty persons on the site, there is no statutory duty to have a first aider – although the hazardous nature of building operations makes it prudent to have someone with a knowledge of first-aid practice present on all sites; in this case a reasonably intelligent, responsible person must be appointed to take charge of any emergency situation and to summon suitable help.

Regardless of the number of people on the site, there must be a first-aid box available which is readily accessible to every first aider and to every employee. Where employees are dispersed over a wide area, it is sensible to provide a number of boxes. The regulations define the items to be included and the minimum quantities of each, depending on the number of employees it serves. These contents must be checked regularly. The box itself should protect the contents from damp and dust and be painted green with an identifying white cross.

Where there are 250 or more person on-site, a suitably equipped and staffed first-aid room must be provided. This is to be in the charge of a fully qualified occupational first aider, and it should be of sufficient size to allow access for a stretcher and to take a couch with space for people to work around it. It must be heated, well lit and ventilated, regularly cleaned and it must not be used for any other purpose.

26.7   Fire precautions

Part of the precautions to be taken to try to prevent the outbreak of fire on sites relates to working equipment and methods, but part of it is concerned with the safety of the provisions made for the health and welfare of the men. Smoking is a well-known cause of many fires and should be prohibited in carpenters' workshops, all storage huts and any storage areas where readily combustible materials are kept. In places where smoking is permitted – mess rooms, changing rooms, canteens, offices etc. – metal ashtrays, partly filled with sand, should be provided and regularly emptied and replenished.

Heating appliances are the next greatest hazard. Portable heaters are easily knocked over or moved to a dangerous position; they should be fixed down to the floor or back to the wall. All heaters and cooking appliances must be properly installed, adequately ventilated (especially to provide combustion air to gas heaters) and, where appropriate, fitted with thermostatic controls.

Metal guards should be fitted over heaters to prevent anything or anyone coming too close and coat stands or drying racks should be fixed to prevent them from being moved nearer to the heater to hasten the drying process but, at the same time, create a risk of the clothes catching alight when dry.

Many huts are raised off the ground to keep them dry. Where this leaves an under-floor space, it should be enclosed with some form of netting to prevent the accumulation of rubbish below the hut and yet still allow the circulation of ventilating air.

An adequate separation between huts should be maintained to

act as fire breaks, but if this is not possible due to the constrictions of the site, the accommodation should be of fire-resistant construction.

Fire extinguishers, of suitable types, should be provided in mess huts and canteens as well as around the site at strategic locations. There are five types of fire extinguisher – water, coloured red; foam, coloured cream, carbon dioxide, coloured black; dry powder, coloured blue; and halon, coloured green. Each is suitable for dealing with specific types of fire only and to use the wrong extinguisher, e.g. a water extinguisher on an electrical fire, can increase the hazard rather than eliminate it.

Part E

Quality control

Chapter 27

Quality standards

27.1 The meaning of quality control

As this complete part is concerned with the subject of quality control, it should start with an examination of what is meant by the term.

In speaking of quality, one implies good quality and that itself is a subjective value. What is thought to be good quality by one person would be totally unacceptable to another, or even by the same person in different circumstances. Suppose a householder builds a brick wall to, say, enclose a heap of garden rubbish. When finished he would, no doubt, be quite pleased with his efforts – the wall would be of acceptable or good quality – even though it may fall far short of the standard capable of achievement by a qualified bricklayer. If, subsequently, that same householder commissioned a builder to construct an extension to his house and the brickwork was of the same quality as his rubbish enclosure, he would have no hesitation in rejecting it, demanding that the walls be pulled down and re-built to an acceptable quality – probably defining the existing house as the standard he wants. Had the householder been a bricklayer he would not, of course, been content with the rough brickwork job for the rubbish enclosure in the first place!

Thus, in considering quality, one is really defining standards of acceptance and these standards can be changed to suit a number of controlling influences such as the cost – in most cases the better the standard the more expensive it is – or the practical level achievable

– there is no point in seeking to get brickwork built to a half a millimetre because the material is not that accurate, but timber can easily be worked to these limits and metals to even finer standards. Similarly, to specify timber to be free from knots is to ask the merchant to find trees without branches, but one can set a standard of quality to control the size and number of knots.

Quality standards also reflect the requirements of use – referring back to the bricklaying householder, what was acceptable as a rubbish enclosure was not acceptable as a house extension.

In the construction industry, the definition of an acceptable standard – and hence the control of quality to reach that standard – is more difficult to achieve than in other manufacturing industries where the product is constantly repeated and the acceptable limits can be closely defined. There is, for instance, the quality of 'acceptable appearance' in construction. It is a characteristic which is impossible to quantify – one cannot measure aesthetics against a standard scale – and, therefore, an extremely difficult quality to control.

There are, however, a number of performance criteria against which the finished product can be judged and the quality assessed. These are:

(a) its ability to achieve the desired durability or reliability;
(b) its projected cost-in-use in terms of running expenses, maintenance costs or replacement costs;
(c) its relevant dimensional precision for the type of material;
(d) its structural stability and strength in relation to its use;
(e) its contribution to the environmental performance of the structure.

For each of these criteria there must be a definition of specification of the standard to be achieved before any control of quality can be instituted. This is the first of five steps in a quality control system.

27.2 Quality control system

The five steps in a quality control system are:

1. set the specification;
2. check the preparations for working;
3. monitor production;
4. correct quality deficiencies;
5. feedback information gained to improve production.

To illustrate these five steps in operation, consider a jobbing builder who has probably never heard of quality control, but who is about to replace the decayed cill and feet of the jambs of a window frame. His first step would be to examine the existing frame to ascertain its construction and from this would conclude that he needs two pieces of softwood, shaped to match the existing frame, for the replacement of the decayed jambs and a hardwood cill of a height which equals the original depth sufficient to give an adequate overhang to the wall and a profile which will suit the glazing or opening casements. This is his specification. He then goes to his suppliers and there selects the timbers he will use on the basis of type, quality and quantity, returns to his workshop, checks that the planes and cutters that he has, with which to shape the timbers, are of a suitable profile and sharp, and that the rest of his equipment is available and in working order. These are his working preparations. Having obtained his materials and armed with suitable tools, he then shapes the timbers in his workshop to follow the carefully measured profiles taken from the existing window, constantly checking that the shapes he is forming are exactly those required, takes them to the site, cuts out the defective timbers and, with care, cuts and fits the new material in place, making sure that no decayed material is left and that the joint between the new and the old is a good, weathertight fit. All this checking and making sure is monitoring the production. He then may find that, despite his careful measurements and skilful shaping, the new jamb sections are not precisely the same as the old and a small amount of adjustment with a chisel and glass paper is needed to ensure that the lines run through correctly.

He is, at this moment, correcting quality deficiencies. The painter may also make some correction of quality deficiencies with stopping in any holes or cracks which should not be there before he starts to paint. Finally, the builder would look at the completed work and congratulate himself on a job well done but, at the same time, question whether the time and trouble taken to patch up the old frame would not have been better spent taking the whole thing out and replacing it with a complete new window. This is feedback information ('experience' in some vocabularies) and can be used to decide, from a more informed basis, whether to repair or replace when faced with a similar task in the future.

27.3 Planning quality control

As already explained, before any form of control can be exercised, there must be a fixed base or point of reference defining the acceptable quality standard and an expressed or implied acceptable deviation from this standard (see section 27.4). Such definitions will be found in the following documents and must be observed:

1. The architect's specification of the workmanship to be employed and the materials to be used. This specification may define the

quality or it may make reference to recognised standards such as British Standards or Codes of Practice.

2. The bills of quantities which specify the desired quality as well as the quantity.
3. The drawings showing what is proposed to be done and, by means of notes, giving information on the quality required.
4. The building contract which links the requirements set out in the above documents into one overall standard.
5. The requirements laid down in British Standard Specifications, British Standard Codes of Practice, where materials or work are stated to be in accordance with these standards, and other relevant national standards such as the National Building Specification.
6. The requirements of the Building Regulations and other relevant legislation, which must be complied with no matter what other standards are set, unless an official relaxation of the requirements has been granted.
7. Samples of the materials as to be supplied, as made up in a sample panel on-site or as in use in an existing building.

Apart from the national and legislative standards, the quality to be aimed for in a building is laid down by the architect on behalf of his client and is arrived at by carefully balancing the best available against the acceptable cost and is the basis of the contract value. Any serious deviation will mean that either the building owner is not getting the value for his money that he should, or he will have to be asked to pay more than he intended (or the contractor will lose some of his profit).

Site management must be made fully aware of the standards of quality laid down by the contract and by law, and this awareness must be communicated to everybody concerned with seeing the work carried out, from the storeman checking the goods in, to the craftsmen erecting the building. It has been known, for instance, for an architect to have approved a brick sample in advance of ordering, inspected and passed the delivered load, approved the laying and pointing of a sample panel showing the finished appearance as intended in the specification, and then for the bricklayer who is actually going to build the wall not to be told of or shown this sample panel nor to be informed about the need to follow the standard set. Clearly in this case, the method or system of monitoring production needs to be examined.

The quality of a finished building derives from a combination of the quality of the materials used and the quality of the workmanship exercised. The materials should be checked each time they are moved to ensure that the initial quality is maintained. It is necessary, therefore, to inspect the goods before they leave the factory to ensure that the product specification has been met; when they arrive on-site to check any damage in transit; when they have been loaded into the store in case there was any damage during unloading; as they are taken out of store to see that they have not deteriorated since delivery; and, finally, just before use to make sure that no loss of quality was suffered between the store and the work place. Thus, it is necessary to arrange for the quality standard to be clearly understood by the inspector in the factory, the deliveryman, the site checker, the storeman, the person in charge of on-site handling and the craftsman.

It should also be recognised that the exercise of quality control cannot be carried out unless time is allowed for it. In the example quoted above, even though the bricklayer was unaware of the sample wall panel, the programme had to be planned to allow for the panel to be built and inspected. Therefore, the brick delivery had to be made at a much earlier date than the day when they would be required for building the wall, to allow for the delivered load to be approved, the sample panel or panels built, inspected and passed or modified. All of which takes time, particularly if progress relies on someone visiting the site to make an inspection

27.4 Deviation from the standard of quality

A drawing shows a wall to be, say, 3.050 m long. When built, the wall turned out to be 3.052 m at one point, 3.051 at another and 3.049 at another and yet, despite the fact that it did not comply with the stated quality of 3.050 m at all points, it was not condemned. This is because it is recognised that variations of one or two millimetres over or under the dimensions of a brick wall are inevitable, because of the nature of the material and its method of production. Such deviation from the standard is accepted and allowed for in the design of abutting components. However, if the wall length and varied between 3.000 and 3.100 m it would certainly have been taken down and re-built because such deviation would be aesthetically intolerable and would exceed the tolerances allowed in the design of other components. There is, then, an acceptable and an unacceptable deviation from the standard laid down which must be observed in controlling the quality of the building.

Whenever any statement of quality or specification of standard is made, it is understood that not every part, component, item or material will conform to an infinite degree. If a board, specified to be a certain thickness, is checked by applying a carpenter's rule to the edge, it may well be found to comply with the specification; but if a particularly keen examiner employs a pair of engineer's callipers to check it, he would find that the thickness is not as specified and would order it to be ground down until it met his standard of measurement. Even then, a test with a micrometer gauge could show that it still did not precisely conform to the standard in all parts and further precision grinding is needed. This process must be continued down to microscopic levels if absolute compliance is demanded.

The principle of acceptable deviation from a declared standard is widely understood, even if not always clearly stated. It can be shown diagrammatically, as in Fig. 27.1, and means that quality control is concerned with identifying which materials or components fall within the acceptable tolerances and which fall outside those limits.

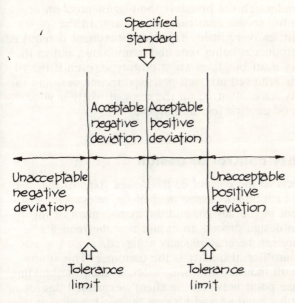

Fig. 27.1 Acceptable deviation

In some specifications, the acceptable deviation is defined as a plus and minus dimension. This is to be particularly noted in components manufactured to fit into a modular co-ordinated building, where everything fits to a grid of planes of reference and must not exceed the modular space between the planes nor fall so short of the plane that the joint to the next component cannot be achieved as intended. Calculation of such allowable deviations take into account manufacturing tolerances, positional tolerances in the placing of the component between the planes of reference, and the maximum and minimum gap which the joint can accommodate.

When the acceptable deviation is not expressed – and it frequently is not – reliance is usually made on normal trade practice. The more accurately a material can be worked, the smaller the acceptable deviations. For instance, it is generally understood that hardwood joinery is much more accurately made with closer joints and finer mouldings than softwood joinery, mainly because greater precision is possible and practical with the close-grained denser hardwood. Furthermore, it is desirable as the finished work will probably be stained and polished to show the beauty of the timber but which will also shown the blemishes and inaccuracies of manufacture, whereas the faults in softwood can often be hidden under a coat of paint.

Reliance on normal trade practice as the basis of quality control is not very precise and is very susceptible to the subjective decisions of the quality controller. This is probably the biggest problem in applying quality control to the construction industry. In the manufacturing industries, very little subjective assessment is required because the same product is being repeated many times, but in the construction industry most buildings are a prototype, even if the means by which it is achieved are well-practised techniques, and the judgement of quality must often be on the basis of whether what has been done is good enough for the purpose.

27.5 The quality control of design

There are two aspects to the control of the design stage of the building process: the continuous assessment of the relevance and quality of the content of the design, and the management of the design team and the design process to ensure that the requisite relevance and quality can be economically achieved.

The ultimate controller of quality is the customer. This is an axiom applicable to all industry and none more so than the building industry. The balance point between the client's expressed desires or perceived needs for his building and his cost limit is the quality level to be achieved. Someone wanting a lot of building for a very small cost must be prepared to accept a low quality of building – many glass and concrete box-like office blocks stand as witness to this approach. If, however, appearance, luxury and visual delight are qualities which the client favours, he must either be prepared to spend more or to reduce the amount of work he has done. It is a fallacy to judge the quality of a building design (and hence the wisdom of the investment) merely on the basis of what it cost to build. Buildings stand for a long time and during that period consume many times the initial cost in running costs and maintenance costs. It is also true that the higher the initial cost – or the better the quality – the lower the total cost-in-use. This total expenditure must be considered when setting quality targets in the control of design.

The basis from which the design team works is the client's brief. Therein is a statement of all the requirements the client sees as being relevant to the proposed building. It will give the number, size and relationships of the various rooms, the standard or standards of finish to be installed, the timing of the work and any phasing needed and a cost limit. To this the design team may add other, technical requirements such as operating systems and servicing requirements to be accommodated.

Constant reference to the brief, accompanied by cost estimations of possible design solutions, is the usual means by which the design team controls the quality of the content. Management of the design team must ensure that all information needed is available when required, that there is an appropriate design team structure which achieves an efficient balance between the various design specialisms and that clear lines of communication between team members are established and maintained.

The actual process of designing has been analysed and broken down into the four phases of assimilation, general study, development and communication. Each of these would be subject to control, both for their content and coverage, and for the time taken. Further information on the analysis of the process of design can be found in *Architectural Design Procedures* of the Longman Technician Series of books.

27.6 Quality control of materials

The control of materials starts – as do most aspects of building – with the architect and the design team. Selection of the most appropriate material to suit the purpose or its location is essential for a satisfactory building and to make any sense of subsequent quality control (there is little point in spending time examining goods for their compliance with an inappropriate quality standard).

Having selected and specified the materials, it is then necessary for either the architect, but more often the contractor, to find a supplier or suppliers who can produce the materials as specified, in the required quantities at the time they are needed for an acceptable price. Given the assurance that these criteria will be met, the order would be placed and delivery awaited.

When the materials arrive on site they should be checked as far as possible before being unloaded. It is more difficult to prove that goods were damaged in transit rather than by mishandling on site once they have been taken off the lorry. If they cannot adequately be checked on the lorry they should be examined immediately they are unloaded to ensure that they conform to the quality specified.

Once past this inspection, the quality of the materials is then a matter of control by the contractor, who must ensure that proper handling and adequate storage facilities are provided as outlined in Chapter 25.

27.7 Quality control of production

The control of the quality of the work in the construction of a building is particularly difficult because of the one-off nature of the operation and because of the lack of any definitive statements on workmanship.

If the person involved is a highly-trained craftsman proud of his skill and opposed to any shoddy work, the need to impose much supervision is not very great – he will be his own quality controller – but it will be necessary to ensure that his good craftsmanship is producing the work shown on the drawings. Therefore, the starting point for control is in the setting-out. This, under most contracts, is the responsibility of the contractor, although many architects will check it for their own peace of mind. Unless the setting-out is correct, nothing will fit properly and, no matter how good the workmanship, the final overall quality of the building will suffer.

Once building work starts, the control of the quality of production is largely a question of continuous supervision; which is why clerks of works are appointed to act on behalf of the architect who cannot spend all his time on the site. All the trades benefit from regular supervision, but certain operations do need particular attention. Concrete work is one; careful observation is required to ensure that the cement mix is obtained with the appropriate water/cement ratio, and that it is properly handled.

Preparing the materials for any of the other wet trades is another area where standards can depart from the specified norm. Frost protection requires vigilance, otherwise good quality workmanship can be spoilt. Painting needs to be frequently checked to ensure adequate coverage of each coat and careful rubbing down between each one, and so on.

Chapter 28

Quality control methods

28.1 Methods of inspection

Materials can be inspected by visual methods, tactile methods and sampling. Each with their respective usefulness and degree of accuracy.

Any lorry load of materials is initially checked visually: the foremen, store keeper or checker runs his eye over the load looking for obvious defects and a practised eye is a valuable and a reliable asset in this respect. The kind of thing he will be looking for is foreign matter such as grass or bits of timber mixed in with hardcore, sand or aggregate; chipped and cracked bricks, blocks, tiles or concrete units; splits, cracks, large knots and blue sap stain in timber; physical damage to prefabricated components etc. Rigorous visual checking can eliminate a very large percentage of sub-standard materials without having to resort to expensive laboratory tests, but care is needed in case the condition of the visible part of the load is not representative of the whole batch.

Tactile inspection, that is feeling the material, has a limited but useful application. An experienced hand can tell the difference between a sound well-seasoned piece of wood and one that is not so good, whereas they both may appear to be the same. It is also possible to tell the age of a bag of cement by its temperature, the strength of a brick by its weight and the sharpness of sand by its grittiness.

Sampling is the most accurate method of inspection and consists of taking samples from widely differing points in the batch and comparing them to the specified standard. This comparison may be by visual inspection, by a simple on-site test, or by sending the samples to a laboratory for analysis.

Where materials carry a manufacturer's certificate or guarantee, a routine visual check is all that is needed, plus an examination of the certificate to see that it is in order and covers all aspects. For instance, some timbers are pre-treated with preservatives before delivery and this carries a treatment certificate; however, the certificate only refers to the treatment and it has nothing to do with the quality, size or quantity of the timber in the load.

28.2 The cost of quality production

If it is assumed that in most production runs of materials or components there will be isolated units which fall short of the quality standard and its acceptable deviation, the cost of reducing or controlling these rejects must be compared to the cost of these failures. In some cases, a high proportion of sub-standard units could lead to the rejection of the whole load because of the cost to the contractor of sorting out the acceptable units.

The cost of maintaining quality can be considered under three headings: failure cost, appraisal costs and prevention costs. Failure costs are made up from the loss of the work which is only fit for scrap, the cost of correcting recoverable rejects, interference with production and the cost of making up the deficiency in the load. Appraisal costs are those connected with the checking that the production of the materials or components is right, while prevention costs are those incurred in trying to make sure that faulty work is not produced in the first place by training of operatives and maintenance and improvement of machines and equipment.

There is no merit in spending money unnecessarily and, therefore, one object of quality control is to achieve the defined standard with the minimum overall cost. All three of the quality costs set out in the last paragraph increase the overall cost of production. The larger the number of rejects, the greater the price of the rest to cover the losses; an increasing percentage of rejects indicates a need for an increasing level of appraisal cost in checking production so these two tend to follow each other. The cost of preventive measures is in an inverse ratio: as more is spent on ways of improving production, less is lost in wastage of results. If, say, the joiner's shop frequently produces frames which are rejected by the on-site inspection because the wood is torn rather than cut, a simple quick overhaul of the machines and sharpening of the cutters used should lead to a significant improvement in quality and a much lower percentage of rejects. There may still be a few due to a variety of causes such as human error, variations in the new material or even a power failure in the workshop. These are more difficult to prevent and, because of the rarity of the event, any preventive measures have a much reduced influence on the level of failures.

Thus, as preventive costs rise, failure and appraisal cost fall. The sum of the two is the overall production cost, and by plotting these on a graph (see Fig. 28.1) the point of minimum total cost can be found by adding the curve for failure and appraisal to that for preventive costs as shown. Not only does this diagram indicate the point of minimum overall cost, it also appears to indicate that any deviation from this point will result in uneconomic operation. It is comprehensible that a reduction in the amount spent on preventive cost in the form of less operative training, more infrequent maintenance, laxity in supervision etc., will lead to more failures and a higher overall cost. What is unexpected is that more stringent preventive measures aiming for even fewer or no defects at all will also increase the overall cost. In other words, to maintain competitive prices, management must be prepared to accept a small percentage of failures.

It is also worth examining the cost of making things in comparison to the value of these things to the customer. While the desired quality is indicated by the client, it is the designer and producer who actually determine the final quality standard and

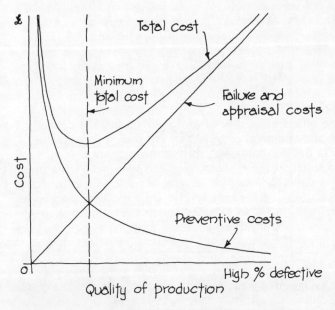

Fig. 28.1 Point of minimum total cost

hence the cost to produce; and as those standards increase so, up to a certain point, does the value to the client.

This can be indicated by another set of cost curves as shown in Fig. 28.2. The horizontal axis indicates the quality of design and manufacture and the vertical axis the cost to produce or the value to the customer. A roughly made article with little design merit still costs a certain amount to produce – material must be bought, wages paid and overheads met – but the end result is not what the client wants and is, therefore, of zero value. A slight increase in quality of raw material, machine setting and maintenance, staff training and supervision can lead to a marked improvement in the product and more interest on the part of the client. Better machines, better qualified staff etc. will lead to even better products but higher costs. Due to higher costs, the client is slightly less interested and his value curve begins to flatten. Eventually, the production cost curve steepens so much that the client is no longer interested at all in paying higher prices for an improved quality which he does not want. As can be seen on the graph, there are two points where the cost to produce curve is above the value to client curve, i.e. the product is not saleable, and an area where the reverse situation applies. The point where the value curve is furthest from the cost curve is the line of maximum profitability.

No account has been taken in this exercise of the effect of production value, nor of market pressures forcing the price below

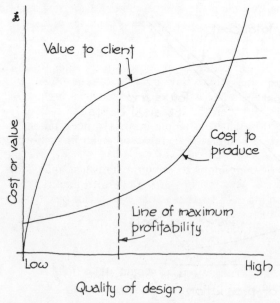

Fig. 28.2 Point of maximum profitability

the maximum profitability line to allow under-cutting of competitors, but it does serve to show the way that control of quality standards and value to client can be related.

28.3 Sampling of materials

As mentioned at the start of this chapter, there are three methods of inspecting materials: visual – by running a practised eye over the load; tactile – by feeling the goods; and by sampling.

The first two rely on the experience of the inspector and, in most cases, this is enough – one does not need a laboratory test to see that a WC pan has been chipped – but with some materials a superficial examination of its appearance will not reveal the information needed to approve or reject it and samples must be taken for closer examination or test.

The idea of sampling is that by examining or applying a test to a small part of the load, one obtains an indication of the quality, strength, cleanliness, grade, proportion of mixture etc. of the complete load. Obviously this only holds true if the sample is representative of the whole load and, therefore, great care must be exercised over how and where the samples are taken. For example, a sample is required of a lorry load of gravel aggregate freshly tipped into the aggregate enclosure. The easiest place to collect some of the material is from the bottom front edge where a small quantity can be scooped up from the concrete floor of the enclosure with little effort. However, as the lorry travelled from the quarry to the site, every bump it went over caused the larger pebbles to settle towards the bottom of the load, and as it was tipped these came rolling out last on to the top of the heap and down its sides to finish up at the bottom edge – just where the sample had been taken. It can be seen then that this sample will probably contain a higher proportion of large pebbles than the load as a whole, and it is to check this very proportion that the sample may have been taken. The correct way to take a sample of aggregate is given in section 28.3.2.

Another purpose of sampling is to obtain a statistical analyis of the precentage of rejects. This can be carried out with materials consisting of a large number of separate units which can be individually checked, such as bricks. Take, for example, a load of 10 000 facing bricks which has arrived on-site and it is obvious that there are a number with damaged faces and arrises knocked off. The supplier would, quite rightly, protest loudly if the load was returned with no further examination and would claim that, although some were damaged at the outer edges, the bulk of the load was satisfactory (and enclosing a bill for transporting 10 000 bricks to and from the site). Alternatively, the contractor would not be very pleased if the site checker spent all day examining and sorting all 10 000 bricks – the supplier would also be unhappy about the length of time his delivery vehicle was held on site. The solution to this problem is to take a statistical sample from all parts of the load to obtain a representative percentage of failures which, if it exceeded the specified quality limit, would then be firm ground for returning the load.

The sample taken should be about 5 per cent of the batch, i.e. the inspector would collect about 500 bricks from all parts of the load. From these he would separate and count all those with damage which makes them unsuitable for use in faced work. This number can then be calculated as a percentage of the sample and compared to the acceptable percentage of damaged bricks. If that acceptable percentage was set at say 5 per cent and thirty bricks of the sample were found to be chipped, then the load would be rejected on the grounds that a statistical sample showed the proportion of damaged bricks to be 6 per cent.

Different tests can be applied to different materials. The following is a selection of the tests applicable to a few of the more common materials. A more extensive survey can be found in *Advanced Building Construction*, Vol 1, in the Longman Technician Series of books.

28.3.1 Inspecting sand

Sand is usually specified to be clean, sharp river or pit sand. The source will be known from the location of the quarry and the cleanliness and sharpness can be initially assessed by rubbing a sample in the palm of the hand. Its grittiness will indicate the sharpness or angularity of the grains and the condition of the inspector's palms will show the cleanliness. If this quick check reveals it to be dirty, a more accurate test can be carried out with a saline solution. To do this, dissolve a 5 ml teaspoonful of sand in 500 ml of water and pour 50 ml of this solution into a glass 200 ml measuring cylinder. Add sufficient sand to bring the level to the 100 ml mark and then pour in more salt water until the level reaches 150 ml. Fit a lid, shake the cylinder vigorously and allow it to stand for three hours. At the end of this time the silt impurities will be visible as a layer on top of the sand, the depth of which should not exceed 6 per cent of the depth of the sand.

Another simple on-site test is to spread a small sample, one grain deep, on a board. Visual inspection will readily reveal the grading of the size of the grains; whether they are too large to fit between the course aggregate of the concrete or to allow bricks to bed properly into mortar, or too small so that they are more like dust and tend to take the place of the cement in a mix.

Further tests can be made to ascertain the nature of the impurities and the character of the grains but, for these purpose, samples must be sent to a laboratory.

28.3.2 Inspecting aggregate

The objects of test on samples of aggregate are similar to those of sand, to see that it is clean and evenly graded. This can be achieved by the methods described for sand.

Grading can be more precisely checked by passing the aggregate through successive sizes of British Standard test sieves. These are made with opening sizes of 75, 63, 37.5, 20, 14, 10, 5, 2.36 and 1.18 mm and 600, 300 and 150 μm. The grading is defined by the percentage of the weight of the sample retained on each sieve.

As explained above, it is important to obtain a representative sample. To do this, ten small samples should be taken at different places in the heap with a scoop of a width not less than four times the maximum particle size. These ten small samples are thoroughly mixed together and then reduced to the quantity required for sampling either by a riffle box or by quartering.

A riffle box (see Fig. 28.3) is a device whereby aggregate poured into the top is separated by transverse fins and each division directs the aggregate to the right and the left alternately, so that half arrives in one collecting box and half in the other. One half is then returned to the heap and the process repeated with the other half until the required sample size is reached.

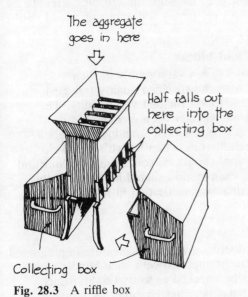

Fig. 28.3 A riffle box

Quartering is the process of dividing a rounded heap of the ten mixed samples into four quarters. Two diagonally-opposite quarters are returned to the heap; the other two are mixed together and quartered again until the sample size is obtained.

If these are to be sent to a testing laboratory, they should be enclosed in a 500-gauge polythene bag, secured with a tie and identified with a label stating the source, the sender and the date. In some cases, a certificate of sampling may be required.

28.3.3 Inspecting concrete

BS 1881 : 1983 describes tests to be carried out on concrete and the apparatus required. The first of these, to check the water/cement ratio and consistency of the mix, is the slump test. This is carried out on samples of the freshly mixed concrete and consists of filling a cone with the mix and measuring how much the mixture drops or slumps when the cone is removed. The method is shown in detail in Fig. 28.4.

The strength of the concrete can only be accurately assessed by testing a cube which has been allowed to set and cure. This cube may be a representative mix of the fine and course aggregates to be used, made up in advance of the main mix, or it may be a works cube, taken from the main mix as it is placed. In either case, the cube in its mould and during transport must be kept covered with wet sacks and stored in water in the laboratory to ensure that it does not dry out more quickly than the main mass, thereby causing its performance characteristics to vary from those of the concrete in use. The usual laboratory test is to measure the crushing strength by placing the cube between two steel plates and applying a load until if fails.

28.3.4 Inspecting bricks and blocks

Bricks, particularly facing bricks, can vary greatly in their appearance and yet comply with the specified quality standard. Indeed, they may have been selected for that very quality. The sampling of damaged bricks has already been mentioned; in addition, they should be struck to test their soundness – a good brick will give off a clear metallic ring when hit with a trowel.

Blocks, whatever their type, should be consistent in colour and true in shape. The object of examination should be to check this and their conformity to the quality specification in respect of type, grade and size.

28.3.5 Inspecting timber

Much timber is now supplied stress-graded and pressure-impregnated with preservative. A certificate is supplied with each bundle of preserved timber and should be checked as to its validity. Each piece of stress-graded timber has been tested either by visual grading and measurement of the size and position of defects or by a grading machine which assesses the bending resistance at regular intervals throughout the length (see Ch. 5, section 5.2). Each piece of timber is appropriately marked with its grade, which should be inspected.

EQUIPMENT

Tamping rod - mild steel
610 mm long × 16 mm dia.

102 mm int. dia.

305 mm high

203 mm int. dia.

Mould

PROCEDURE

Quarter fill mould with concrete.
Tamp 25 times with steel rod.
Repeat three more times until
the mould is full.
This filling must be completed
in 2 minutes.
Immediately it is full lift the mould
vertically and place it beside
concrete sample.
Lay the tamping rod across top of
mould and over concrete to see
how much it has slumped.
Measure the distance between
rod and the top of the concrete
to obtain the slump:

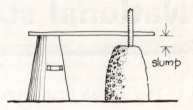

slump

MAXIMUM SLUMP VALUES

Type of concrete	Slump (mm)
Vibrated very high strength	0
Vibrated high strength and mass concrete	0-25
Hand compacted mass concrete	25-50
Vibrated normally reinforced	25-50
Vibrated heavily reinforced	50-100
Hand compacted normally reinforced	50-100
Hand compacted heavily reinforced	100-150

Fig. 28.4 The slump test

In addition, the timber should be inspected for any of the following common defects:

SHAKES splits running the length of the timber
WANE angles rounded off following the surface of the log
BLUE STAIN a fungal infection of the sap wood of softwoods

Chapter 29

National standards

29.1 The British Standards Institution (BSI)

The British Standards Institution is the national body within the
United Kingdom responsible for setting standards. It is an
independent organisation governed by an Executive Board composed
of elected representatives from each section of industry and
nominees from the Trades Union Congress, the Confederation of
British Industry, the nationalised industries, the Department of
Trade and Industry, the Ministry of Defence and the Department of
the Environment. Responsibility for the work of standardisation is
delegated by the Board to Divisional Councils each concerned with
different sectors of industry. Building standards are the responsibility
of the Council for Building and Civil Engineering Works. Working
under the Councils are the Industry Standards Committees which
deal with more detailed sectors of industry.

The BSI is non-profit making and gains much of its finance from
the subscriptions of its members – companies and firms, trade
associations, local authorities, professional institutions, private
individuals and any other body interested in its work. It also earns
money from its publications and its charges for services such as
certification marking and it receives a government grant which is
proportional to its subscription income.

Its formal function is achieving and recording agreement on
known techniques, the process of which sometimes shows gaps in
present knowledge or leads to further development work. As the

BSI is not a research body, this work must be undertaken by others, often the Building Research Establishment and, until the research or development is completed, the available information may be published as a Draft for Development (prefixed DD).

The function of the BSI is described in its Royal Charter of 1929 as being:

(a) To co-ordinate the efforts of producers and users for the improvement standardisation and simplification of engineering and industrial materials so as to simplify production and distribution, and to eliminate the national waste of time and materials involved in the production of an unnecessary variety of patterns and prices of articles for one and the same purpose.
(b) To set up standards of quality and dimensions, and prepare and promote the general adoption of British Standard specifications and schedules in connection therewith. . . .

The specific reference to engineering in the above quotation is due to the BSI having started as the Engineering Standards Committee solely concerned with that one industry. The original meeting took place in 1901 and was chaired by Sir John Wolfe Barry and attended by representatives from the Institute of Civil Engineers, the Institute of Mechanical Engineers, the Institute of Naval Architects and the Iron and Steel Federation. In 1918 the Committee became the British Engineering Standards Association and eventually, in 1931, the British Standard Institute.

As examples of the achievement of the stated aim of simplification and elimination of wasteful variety, the first Standard Specification, now BS4, brought down the number of structural steel sections produced from 175 to 113 and the second, now obsolete, concerned with tramway rails, reduced the variety of sizes and shapes produced by all the different steel mills from 75 to 5.

29.2 Standard specifications

There are over 1500 BSI publications affecting the construction industry, mostly either Standard Specifications or Codes of Practice. About 300 of these deal with the plant and equipment, which is the direct concern of a contractor, the other 1200 or so cover a range of subjects and contain information essential to everyone working in the building industry.

They are important to the architect and other members of the design team because they are a comprehensive statement of the properties required in a product to ensure acceptable quality, performance and compatibility with other components. They also give test methods to verify that these requirements are met. The building contractor also finds it important to be aware of and

304

constantly to refer to the standards to check that his work complies with the architect's specification and what is considered to be the best and latest building techniques.

Compliance with British Standards is, unless enforced by other legislation such as the Building Regulations, purely voluntary, but goods marked with the familiar Kitemark (see Fig. 29.1) of the British Standards Institution carry with them an assurance of quality. Not all Standards are concerned with quality; some lay down standard sizes, others define test procedures, others deal with the working methods or procedures etc. In quoting, or requiring adherence to, a particular Standard, the number quoted should be checked and the content studied to make sure that that particular Standard does cover the subject and the particular aspect required. For instance, BS 644 deals with wood windows, but if the matter for concern is the testing of windows, then either BS 5368 or BS 4315 are needed. However, if the subject is the quality of the timber of which the window is made, or its workmanship, the BS 1186 is the reference required, and so on.

Fig. 29.1 British Standards Institution registered certification trade mark (Kitemark)

A specification that a product is to comply with one British Standard may bring a whole host of others into operation because standardisation goes right back to basic materials. Thus BS 5911: Part 1: 1981, which has as its subject concrete drainpipes and fittings, refers to twenty-five other standards for cement, aggregate reinforcement, admixtures, joint rings, test methods and building drainage generally. These Standards, in their turn, require compliance with other Standards which may extend outside the area served by the Council for Building and Civil Engineering Works.

BS Handbook 3, *Summaries of British Standards for Building,* is a four-volume work which tells architects and builders all they need to know about individual standards, leaving out only the methods of test and production procedures which are more the concern of product manufacturers. It summarises over 1200 Standards and is in a loose-leaf format which is updated annually.

29.3 British Standard Codes of Practice

In most cases, it is not enough just to ensure that materials are of a good standard; the way they are used and assembled must be of an equal quality if the finished work is to achieve the desired results. For this purpose, codes of practice have been produced. They take the form of recommended working methods which, if followed, produce sound practice; but compliance with these recommendations does not remove the necessity to satisfy the requirements of any legal enactment,

The origination of Codes of Practice was in 1942 when, in September of that year, the Council for Codes of Practice was formed under, what was then, the Ministry of Works. This Council was made up from members nominated from fifteen professional and scientific institutions concerned with building, plus representatives of the British Standards Institution and the government. The Codes were issued by the BSI on behalf of the Council and were defined as:

recommendations for good practice to be followed during design, manufacture, construction, installation and maintenance, with a view to safety, quality, economy and fitness for purpose.

Much of the work of Codes of Practice has been absorbed by the BSI and many publications which originally were Codes of Practice (prefixed CP) are now British Standards, as are new Codes of practice.

29.4 Certification

As already mentioned, compliance with a British Standard is purely voluntary, unless specific reference is made in legislation. A policy of compliance may be adopted by the manufacturer or it may be imposed by the purchaser looking for an assurance of quality. If an article is marked with a relevant British Standard number, the manufacturer is claiming that his product complies with that in all respects. Failure to do so can mean that the manufacturer is guilty of misrepresentation, a crime dealt with under other relevant legislation. The kitemark shown in Fig. 29.1 is the BSI's registered certification trade mark and is an assurance of quality. To be able to display this, the manufacturer must register with the BSI who certify that the goods have been produced under a system of supervision, control and testing operated during manufacture and including periodical unannounced inspection of the manufacturer's works by BSI inspectors.

Testing of products in connection with certification is done at the BSI Test Centre at Hemel Hempstead. The primary concern of the Test Centre is this testing, but its services are available for other

testing work on a commercial basis, and provides an inspectorate
and test service for other organisations concerned with quality
control.

29.5 The British Board of Agrément

The building industry, as any other, is the scene of constant
technological development and innovation, with new products
continuously coming into the market. Many of these new products
cannot comply with a British Standard since none exists and, being
new, they do not possess a proven record of success. Since the
building industry deals with a final product which is intended to last
a very long time, there exists a deep mistrust of new, untried or
tested products which may lead to early and costly failures of the
building, and yet all sectors of the industry are constantly looking
for reliable ways to improve their standards or reduce their costs.

It is really to help to overcome this paradox of resistance to new
ideas that the British Board of Agrément has been set up. It was
originally known as the Agrément Board but has now changed its
name. The BBA is principally concerned with the testing,
assessment and certification of specific products for the construction
industry, in order to secure the ready acceptance of the products
concerned and to ensure their safe and effective use. The subjects
for assessment are normally new or innovatory products, systems or
techniques, but existing ones may be re-assessed should a new use
arise or should there be a change in circumstances relating to the
product such as, for example, a revision of the Building
Regulations.

It was in 1963 that, at a meeting arranged by the Royal Institute
of British Architects, the subject of the Agrément System then
operating in France was first discussed. This led to the appointment
of a Committee on Agrément by the Ministry of Public Building
and Works under the chairmanship of Sir Donald Gibson, the
Director General of Research and Development. The committee
reported its findings in 1965, which recognised the need to establish
a national authority for the testing and assessment of new building
methods and materials to give authoritative assurance that these
innovations met the claim made for them. Following receipt of this
report, the Minister set up the Agrément Board, as a company
limited by guarantee, on 2 May 1966.

The present Board is sponsored by the Department of the
Environment and its Chairman and Council members are appointed
by the Minister for Housing and Construction. It is a member of the
European Union of Agrément (UEAtc) whose objective is to
facilitate international trade in the building sector. UEAtc was

created to minimise the need for one country to retest imports from another.

Unlike British Standards, which deal with building products as types, i.e. timber windows, steel reinforcement, clay pipes etc., the British Board of Agrément Certificate refers to a specific named product from one manufacturer. It is obtained by that manufacturer making an initial approach to the BBA with a request for his product to be assessed. He must then supply preliminary information on his product, including method of manufacture, chemical composition, proposed use, details of procedure for installation etc. From this a programme of work and the amount of the fee are worked out. On payment of the fee, samples of the product are tested – either by this Board or by laboratories acting as agents; the relevance of the product to legislation, such as the Building Regulations, is examined; the firm's quality control arrangements are inspected and visits are made to buildings or trial areas where the product is in use. If all these are satisfactory, a Certificate is issued.

The Certificate is usually in three parts. Part I Certification states the product to which the Certificate relates, who makes it and who markets it, its use, an assessment of its performance and its ability to meet Building Regulation requirements and any conditions attached to the certification – such as the need to maintain standard of quality or the need to carry out periodic inspections. Part II gives a technical specification and relevant design data and Part III sets out the technical investigations and tests carried out. It is signed by the Board's director and is normally valid for three years but may be renewed for a further three years or more.

A purpose of the Certificate is to cover the period between a new product or process evolving, and the production of a suitable British Standard to cover it. Normally, the BBA do not issue Certificates for products already covered by a British Standard. To control any possible overlap of work and to ensure that the issue of a certificate leads eventually to a new British Standard, the BBA and the BSI work closely together and the role of each was defined in a formal agreement in 1973. A BSI/BBA committee meets regularly to develop the policy of co-operation.

Certification of products is not the only assessment carried out by the Board. In many cases, a satisfactory product can be badly installed and its value is lost or, worse still, it becomes the source of other building troubles. To make certain that the certified products do perform as the Board says they should, the firms carrying out the work must be assessed, approved as installers of the system or product and registered under the Board's Approved Installation Contracts Scheme. Thus, satisfaction can only be assured for, say, cavity wall insulation injection if the insulating material carries a

308

Certificate and the firm carrying out the work is on the list of
approved installers. To make certain that the standards of
workmanship and skill which the installers showed when assessed
are maintained, the Board's inspectors make unannounced
surveillance visits during the period of the validity of the
registration.

Bibliography

G. Barnbrook, *House Foundations*. Cement and Concrete
Association
A. J. Brook, *The Cladding of Buildings*. Longman
R. Chudley, *Building Finishes, Fittings and Domestic Services*.
  Longman
H. A. Taylor, *Environmental Science*. Van Nostrand Reinhold
D. W. Durrant, *Interior Lighting Design*, 4th ed. Lighting Industry
  Federation Ltd
Code of Practice, *Means of Escape in Case of Fire*. Greater London
  Council
E. G. Butcher and A. C. Parnell, *Smoke Control in Fire Safety
  Design*. E. & F. N. Spon
Trees and Buildings. RIBA Publications Ltd and The Tree Council
Design Guide for Residential Areas. County Council of Essex
The Building Employers Confederation, *Construction Safety*. The
  Building Advisory Service
David Croney, *The Design and Performance of Road Pavements*.
  HMSO
A. A. Lilley and A. J. Clark, *Concrete Block Paving*. Cement and
  Concrete Association
Dept. of the Environment, Dept. of Transport, *Design Bulletin 32,
  Residential Roads and Footpaths*. HMSO
John Uff, *Construction Law*. Sweet and Maxwell
Speight and Stone (eds), *A. J. Legal Handbook*. The Architectural
  Press
Ian E. Chandler, *Materials Management on Building Sites*. The
  Construction Press
R. H. Caplen, *A Practical Approach to Quality Control*. Business
  Books
House's Guide to the Construction Industry. House Information
  Services Ltd

Index